Studies in Applied Philosophy, Epistemology and Rational Ethics

Volume 56

Studies in Applied Philosophy, Epistemology and Rational Ethics (SAPERE) publishes new developments and advances in all the fields of philosophy, epistemology, and ethics, bringing them together with a cluster of scientific disciplines and technological outcomes: ranging from computer science to life sciences, from economics, law, and education to engineering, logic, and mathematics, from medicine to physics, human sciences, and politics. The series aims at covering all the challenging philosophical and ethical themes of contemporary society, making them appropriately applicable to contemporary theoretical and practical problems, impasses, controversies, and conflicts. Our scientific and technological era has offered "new" topics to all areas of philosophy and ethics – for instance concerning scientific rationality, creativity, human and artificial intelligence, social and folk epistemology, ordinary reasoning, cognitive niches and cultural evolution, ecological crisis, ecologically situated rationality, consciousness, freedom and responsibility, human identity and uniqueness, cooperation, altruism, intersubjectivity and empathy, spirituality, violence. The impact of such topics has been mainly undermined by contemporary cultural settings, whereas they should increase the demand of interdisciplinary applied knowledge and fresh and original understanding. In turn, traditional philosophical and ethical themes have been profoundly affected and transformed as well: they should be further examined as embedded and applied within their scientific and technological environments so to update their received and often old-fashioned disciplinary treatment and appeal. Applying philosophy individuates therefore a new research commitment for the 21st century, focused on the main problems of recent methodological, logical, epistemological, and cognitive aspects of modeling activities employed both in intellectual and scientific discovery, and in technological innovation, including the computational tools intertwined with such practices, to understand them in a wide and integrated perspective. **Studies in Applied Philosophy, Epistemology and Rational Ethics** means to demonstrate the contemporary practical relevance of this novel philosophical approach and thus to provide a home for monographs, lecture notes, selected contributions from specialized conferences and workshops as well as selected Ph.D. theses. The series welcomes contributions from philosophers as well as from scientists, engineers, and intellectuals interested in showing how applying philosophy can increase knowledge about our current world. Initial proposals can be sent to the Editor-in-Chief, Prof. Lorenzo Magnani, lmagnani@unipv.it:

- A short synopsis of the work or the introduction chapter
- The proposed Table of Contents
- The CV of the lead author(s).

For more information, please contact the Editor-in-Chief at lmagnani@unipv.it.

Indexed by SCOPUS, ISI and Springerlink. The books of the series are submitted for indexing to Web of Science.

More information about this series at http://www.springer.com/series/10087

Tommaso Bertolotti
Editor

Cognition in 3E: Emergent, Embodied, Extended

Multidisciplinary Perspectives

Springer

Editor
Tommaso Bertolotti
Department of Humanities
Philosophy Section
University of Pavia
Pavia, Italy

ISSN 2192-6255 ISSN 2192-6263 (electronic)
Studies in Applied Philosophy, Epistemology and Rational Ethics
ISBN 978-3-030-46338-0 ISBN 978-3-030-46339-7 (eBook)
https://doi.org/10.1007/978-3-030-46339-7

This Springer imprint is published by the registered company Springer Nature Switzerland AG
The registered company address is: Gewerbestrasse 11, 6330 Cham, Switzerland

Preface

There are more things between heaven and earth, dear reader, than dreamt of in your philosophy. This is the warning that William Shakespeare inscribed onto the lips of Hamlet, the prince of Denmark. Hamlet was rebuking Horatio and his incredulity before the ghost of the dead king.

In the Renaissance, "philosophy" designated a wider array of matters than the word does today: it encompassed natural philosophy. Galileo and the Bard of Avon were presumably born the same year (1564): Science in its contemporary understanding was still to be born, and philosophy was the main repository of true knowledge about the world. For the sake of this argument, it should be noted that Science and Philosophy share a very similar etymology: the former is the love of knowledge, while the second means nothing but Knowledge—in Latin, *Scientia* from the verb *Scire* which simply means "to know".

What is then Hamlet telling Horatio? That there are more things between heaven and earth than in his knowledge. Is this a but flourished way to say, "There are things that you don't know"? Yes and no. That famous sentence does include the essence of Shakespeare's theatre.

Knowledge and ignorance are the main characters in all of Shakespeare's plays. The plot unraveling across the acts is indeed a function of knowledge. A character ignores the true identity of another, or is perchance even oblivious of her own nature, or is mistaken in assuming the reaction of a third character to an event, and so on and so forth. Shakespeare made it clear that stories are not shaped by what character *do*, but by what they *know*. What characters know triggers actions intended at helping them pursue their goals. As they pursue their goals they travel, and as they travel they gather new knowledge, which prompts different courses of action. The resolution of a play does not signify that everything that was to be done was done indeed, but that everything that was to be known was eventually known and subsequent actions are, from that point of view, immaterial.

In Shakespeare's plays, narrative resolutions are first of all cognitive resolutions. Cognition can be defined as the capacity to display adaptive behaviors, that is to say to react in a smart, non-automatic way to changes in one's environment. Cognitive resolutions obtain when characters know everything they need to know in order to behave in a smart way in their environment—which is also composed of other characters. Death, a common fate for tragic characters, is also a cognitive resolution, that is the end of the cognizant. Once dead, there is no more behaving, so the quest for knowledge ends.

The death of the tragic characters is usually accompanied by a reveal for the spectator: the spectator, or the reader, is like a daemon who gathers all of the relevant knowledge from all parties at stake, and would know what portion to assign each to each of them so they can each (or almost) triumph in their plan. Had Romeo known, upon arriving at the Capulets' monument, that Juliet was asleep and not dead, in all likelihood he would have opted for a more adaptive choice than ending his own life. No one told him to take a step back from the situation that was unfolding before his eyes. No one suggested that there could be more things between heaven and earth than dreamt of in his philosophy: for instance, a plot involving sleeping drugs devised by a friend in order to crown their dream of love.

How, then, can we understand the famous warning that Shakespeare spoke through Hamlet's royal lips? Again, it's not so much about the things that you know or do not know. It's rather about how what matters is not necessarily only in your mind. It's an invitation to search for the parts and processes leading to your cognitive resolution "between heaven and earth", where most things are, and not just "in your philosophy".

To limit the exercise of one's cognitive abilities to what's within our current knowledge, without plugging into the world (to use Andy Clark's expression), is hardly ever a rewarding choice; if only because, according to the recent views of embodied and extended cognition advocated for in this volume, there is so much more cognition than what goes on in our mind. Provided it makes sense to think that what goes on in our mind can be isolated from all the rest that is "between heaven and earth".

Hamlet is often portrayed addressing a skull as he asks the famous question whether "to be or not to be". This book could have been conceived out of a similar iconography, each author asking the skull, "where do you plug into?"

Even accepting the broad definition of cognition as the *capability to display adaptive behavior*, much remains to explore and wonder about, namely, where are the roots of this capability. These are the three E addressed in the volume: Emergent, Embodied, and Extended. Focusing on cognition as an emergent property means to investigate how the processes at stake can be drawn back to simpler levels and deep relationships that go beyond (and beneath) what we usually understand as "being cognitive". To study the embodiment means to focus on how much cognitive properties depend on the kind of bodies they are nested into. The extended paradigm is a conceptually natural development of embodied theories, as it explores cognition as it happens in the interactivity (and interpassivity) between the different cognitive subjects and cognitive objects/supports spanning across space and time.

This book was born from a workshop I organized at the University of Pavia, Italy, in May 2018. In turn, that workshop was born by chance out of another extension and mingling of different cognitions around a conference in London, in

November 2017. We don't really go to conferences for talks, but to talk between talks. The conversations we had in December with Tomie Hahn, Chris Mays, and J. Scott Jordan deserved and reclaimed a proper occasion, the perfect stage as far as emergence, embodiment, and extension are concerned. The moveable feast was convened in Pavia six months after our first encounter, in a setting that could afford extensive presentation and the thoughtful exchange they sparked. Tomie, Curtis, Chris, and Scott were matched by my local partners in research: Lorenzo Magnani, Selene Arfini, and Marco Viola—whose research interests perfectly suited this lineup.

The results of our conversations, twice distilled, are in this volume: just abstracts, not papers, were required before the workshop to let each mind be exposed to each other's research and let the final words reflect our mutual exposure to different, yet compatible perspectives.

The workshop was funded through the Blue Sky Research Project (2017–2019) I was awarded by the University of Pavia. I want to express my gratitude to the Department of Humanities—Philosophy Section for supporting me in the organization. This workshop could not have been possible without the teachings of Lorenzo Magnani and the support of Selene Arfini. I am deeply indebted toward the Department of Economics for letting us take advantage of their beautiful conference hall: I address a special thanks to the director, Prof. Maiocchi, and to Lorenza Magnani for facilitating the connection. Special kudos to the admins of the Department of Humanities for handling and solving the complexities that international invitations still caused in 2018. I was moved by and grateful for the involvement of my students (from the class of cognitive philosophy) into the workshop, especially the banding session organized by Tomie Hahn. We had an amazing ride, which I daresay is further proved by the quality of the contributions to this volume. Without further ado, let us turn to what the participants had to say about Cognition in three E's: emergent, embodied, and extended.

Los Angeles, USA
December 2019

Tommaso Bertolotti

Contents

Distributed Cognition in Aid of Interdisciplinary Collaborations

Selene Arfini

Abstract What does it take to perform collaborative interdisciplinarity with good epistemic and academic results? This pragmatical question, slightly rephrased, has been one of the few key issues of the philosophical studies on interdisciplinarity since the Seventies. In this paper I aim at addressing that question adopting a conceptual framework weirdly not yet used for this purpose: distributed cognition theories. In particular I will focus on the embodied, emerging, and extended nature of cognitive activities at the core of successful examples of collaborative interdisciplinarity. In the first section of this paper I will briefly review the literature on interdisciplinary collaborations. In the second section I will present the perspective from which I aim at addressing their recurrent problems: a broadly conceived distributed cognition theory, which incorporates insights from the extended mind approach, and from the theories on emerging and embodied cognition. In the third section of the paper I will analyze some of the usual emerging problems of collaborative interdisciplinarity by referring to some well-documented case studies. Then, I will propose some ways to face those problems in the organization and development of a collaborative interdisciplinary project, referring to it as a complex system of distributed cognitive activities.

1 What Does It Take to Perform Good Interdisciplinary Collaborations?

"What does it take to perform collaborative interdisciplinarity with good epistemic and academic results?" The first time I addressed this question was in 2016 and I was beginning a visiting period at Tilburg Research Center for Logic and Philosophy of Science. I won that position because I asked that daring question and I thought I had a great plan to answer it. I began my research and I published the abstract and the first draft of this article on the page of my profile in my university website. At that time I was at the second year of my Ph.D. course in a small university, so by

---

S. Arfini (✉)
Department of Humanities, Philosophy Section, University of Pavia, Pavia, Italy
e-mail: selene.arfini@unipv.it

T. Bertolotti (ed.), *Cognition in 3E: Emergent, Embodied, Extended*,
Studies in Applied Philosophy, Epistemology and Rational Ethics 56,
https://doi.org/10.1007/978-3-030-46339-7_1

publishing the draft of this paper on that platform I did not really expect feedbacks on it, but I mainly wanted to give myself a push to complete my to-write list. During my period at Tilburg, I got close to finish the paper but I ended up putting aside the project for a better time. Notwithstanding the low visibility of that article—not even finished—during the past three years I received at least six emails of scholars who wanted to start an interdisciplinary project and were looking for operative advices on how it can be done successfully. Since it was a never-really-published (and almost never finished) paper I give myself two reasons for that unusual luck.

First of all, since high specialization has become the leading trend for academic institutions, interdisciplinarity is the solution that allows researchers of well-defined disciplines to still aim high and reach beyond their disciplinary boundaries. Second, there is no consensus on what it takes to perform collaborative interdisciplinarity with good epistemic and academic results and there are no manuals one can follow to create a good interdisciplinary team. These two reasons are confirmed by the contemporary literature on the topic (which I will briefly review in the next section), which generally describes collaborative interdisciplinarity as a highly desirable *mess* (borrowing the unorthodox term from Ackoff (1974a, b), Donaldson et al. (2010)).

For these two reasons, I believe that the pragmatical question regarding collaborative interdisciplinarity is still worth asking. The main problem of addressing that question is doing it in a way that can drive forward the research regarding interdisciplinarity studies without stumbling in the same inconclusive steps that were already walked by previous literature. Hence, in the first section of this paper I will briefly review the literature on interdisciplinarity and interdisciplinary collaborations. In the second section I will present the perspective from which I aim at addressing the problems of collaborative interdisciplinarity: a broadly conceived distributed cognition theory, which incorporates insights from the extended mind approach, and from the theories on emerging and embodied cognition. In the third section of the paper I will analyze some of the usually emerging problems of collaborative interdisciplinarity by referring to some well-documented case studies. Then, in the last three subsections of the paper I will propose some ways to face those problems in the organization and development of a collaborative interdisciplinary project, thought as a complex system of distributed cognitive activities.

Hence, let us begin by discussing why collaborative interdisciplinarity is considered a *messy* business in academic and scientific frameworks alike.[1]

---

[1]Here I need to add a necessary terminological note. In this paper I will use "interdisciplinarity" without specifying the difference between it and similar terms—such as transdisciplinary, multidisciplinary, and others—that have been painstakingly described and differentiated by a part of the philosophical literature in the last forty years or so—cf. Jantsch (1970), Klein (1990), Thorén and Persson (2013). I will not adopt a high differentiated distinction between inter-, multi-, and trans-disciplinarity for two reasons: (1) the main difference between interdisciplinarity, multidisciplinarity, and transdisciplinarity revolves on the connections between different fields rather than between people. Since I'm more interested in the forms of collaboration that are established between researchers with different background knowledge, I believe that specifying the inter-, multi-, transdisciplinary relations between fields would be quite gratuitous in this research. (2) I will not adopt interdisciplinarity as a precise term to define "coordination by high-level concepts" as described by Jantsch (1970), nor I will adopt a more specific philosophical description for it (as the one

2 The Interesting Cognitive *Mess* of Collaborative Interdisciplinarity

Collaborative interdisciplinary is widely known to be an important as much as a daring proposal in both scientific and academic frameworks. It is important for several reasons.

First, it allows to address challenging problems that cannot be solved remaining in the strict perspective of a singular discipline, such as environmental changes, societal issues, and the impact of new technologies (according to Donaldson et al. (2010, p. 1521) interdisciplinarity is not only a *messy* business but it is also a response to *mess*, as it helps addressing "irreducibly complex problems of the world").

Second, when successfully achieved, it represents an academic valuable result. For example, Larivière et al. (2015) found a consistent correlation between interdisciplinary publications and a higher level of citation, especially on the long period (Lachance and Larivière 2014). More than that, the openness to interdisciplinary collaborations is favorably mentioned in the mission statements of many research funding agencies, such as the National Science Foundation, the National Academy of Sciences, and the European Union Research Advisory Board (Maki and MacLeod 2016; Maki 2016).

As a third reason to look at interdisciplinarity as something to aim at achieving, in scientific practice it is actually growing (in different forms) and now many philosophers of science look at the study of interdisciplinary as a field of new and exciting research.[2]

Nevertheless, even if the phenomenon of interdisciplinary collaborations and the studies regarding it are growing (indeed, it has been growing since the Seventies—cf. Jantsch 1970), there are problematic traits that keep this practice a troublesome thing to conceive, organize, and achieve.

First, as already mentioned, there is still no consensus on what it takes to perform successful interdisciplinary research or what constitutes a good form of interdisciplinary collaboration—as already denounced by Weingart (2000), Holbrook (2013), Maki (2016), Koskinen and Maki (2016). Descriptions of good interdisciplinary relations go from very demanding (or *hyperformalized* in the objectives of research fundings, according to Strathern (2004)) to almost morally loaded. Indeed, it is

---

provided by Apostel et al. (1972) as activities involving the integration of concepts, procedures, epistemology, terminology, and data). Indeed, I will use interdisciplinarity in the same way that the non-philosophical literature uses it, as the umbrella term that covers every kind of interaction between researchers with different disciplinary expertise. Anyhow, I am aware that the interaction between these people can take many forms, so I will specify in every occasion, that I will discuss "collaborative" interdisciplinarity, as the interaction among people from different fields who aim at collaborate to reach a common goal. In this sense, my use of the word is closer to the definition provided by Boden (1997, p. 18) of *co-operative interdisciplinarity*, as "an enterprise in which several groups with complimentary skills work towards a common goal, actively co-operating on the way."

[2]For a more specific view on the philosophical aspects of philosophy of interdisciplinary science see Maki (2016).

not surprising to read passages like these ones in the literature: "interdisciplinary collaboration requires *great friendship*" (Buller 2009; Maki and MacLeod 2016); "Successful interdisciplinary research demands good science as much as it demands personal values related to *patience, trust, responsibility*, and *honesty*, all attributes related to being a *respected citizens*" (Naiman et al. 1998; Naiman 1999, italics added for emphasis).

Moreover, even if it is true that many funding agencies stress the value of interdisciplinary research in their mission statements, different studies describe a different, less optimistic picture of the success rate of interdisciplinary research as far as fundings go. For example, biases toward disciplinary studies are well known to affect research funding (Metzger and Zare 1999), and Bromham et al. (2016) showed that interdisciplinary proposals have consistently lower funding success rates.

If we surely can find many specific causes for that, it is relatively easy to argue that it mainly derives from the fact that, as already argued by Donaldson et al. (2010, p. 1525), collaborative interdisciplinarity is a "messy business, which can also incur in unexpected costs, given the need to invest significant time in building collaborative relationships, developing a shared language and honing a common perspective from disparate viewpoints."

Notwithstanding the pros and cons of interdisciplinarity, however, asking whether it is worth pursuing it—or, less pragmatically important, if it is a good topic of research in philosophy of science—is no longer a question we should consider. Indeed, it appears that disciplinary topics are less optional than twenty years ago. First and foremost because it is not a case that attempts at initiating interdisciplinary research are becoming more common as hyper-specialization is becoming a vast phenomenon in both academic and scientific fields. The now well-known phenomenon of growing specialization, as much as it comes with the advantage of providing more advanced and specific knowledge for any subfield of research, it also narrows down the potential theoretical and pragmatical goals of each specialistic area. To aim at complex, "hybrid" and multifaceted problems (Maki and MacLeod 2016), various subfields of specialization need to be put together and researchers of various background knowledge need to be instructed on other disciplines' reach and expertise. Interdisciplinarity, in various forms, is now a common *necessity* of researchers, who need at least to read studies from outside their disciplines in order to keep their work valuable, updated, and, ironically, funded.

So, asking what it takes to perform collaborative interdisciplinarity with good epistemic and academic results is still a question that needs a sensible answer. The problems that derive from asking it are mainly two and can be summed up with two sub-questions: (1) what do exactly "good epistemic and academic results" entail? And how could we provide a good answer for this question that has not been proposed and unsuccessfully tried out yet? In the next subsection I will answer the first query, and in the next section I will provide reasons for adopting the conceptual framework of distributed cognition to address pains and gains of interdisciplinary collaborations.

2.1 The Two Main Problems of Interdisciplinary Collaborations

First let's clear something out. If two researchers organize each an interdisciplinary project that aims respectively at achieving good epistemic results and good academic results, that probably means that they have two very different projects in mind. Any person who has at any point worked in an academic environment knows that not every epistemic result is a good academic result and vice-versa, and trying to get them both while aiming at either one or the other is not something that comes naturally.

Good academic results entail publications, fundings, and having a researcher profile that is highly competitive in an environment that, as already mentioned, has become extremely specialized. More or less these three things support each other: the more publications you have in a narrow field of research, the more fundings you are likely to obtain, the more competitive your profile gets with time (this wheel goes on and on, incrementing the infamous Matthew's Effect in academia (Merton 1968)). So, remaining a researcher interested in just one's own disciplinary topics would require effort (given the emerging necessity of having, at least, interdisciplinary interests), but it would be safer than adventuring in interdisciplinary areas, especially if you are a young researcher. Indeed, since the benefits of interdisciplinary research can be usually appreciated with time and interdisciplinarity poses some challenges that researchers are not usually prepared to face in disciplinary training, beginning an interdisciplinary project is usually safer if you are an already experienced researcher who has not the same urgency to publish or getting fresh fundings as a younger one. For instance, as I previously commented, it is true that interdisciplinarity brings more citations (so good thing for publishing—even if in the long run) but at the same times it is difficult to find good venues to publish a paper that explores interdisciplinary topics, especially if you have undertook the daring project of putting together a team from disciplinary backgrounds that are not traditionally connected.

Moreover, it is fairly common that research results take longer to publish when prepared in an interdisciplinary group, since there are different methodologies and styles to take into consideration (Naiman 1999; Thagard 2006; Andersen 2013; Paletz et al. 2016). At the same time, it is true that interdisciplinarity is always mentioned in the instructions for grant applications, but finding members for the academic jury that have the expertise to fairly judge an interdisciplinary proposal is not the easiest thing to do and, indeed, it seldom happens. Moreover, embarking in an interdisciplinary project naturally requires a loss in specialistic updated research as confirmed by Andersen (2013), Donaldson et al. (2010), Buller (2009). So, from a purely academic point of view, interdisciplinarity can be described as a bold move that brings results in the long run, and that can easily backfire sooner, especially for young researchers.

Instead, from a merely epistemic point of view, interdisciplinarity is a stand alone goal when discussing complex issues that can be appropriately faced by a plurality of views. As already argued by Maki and MacLeod (2016), different fields are full of hybrid problems that should be addressed by experts from many background areas and academic disciplines (such as environmental science, cognitive science, biology, psychology, not to mention philosophy and its constant struggle to remain relevant in a scientifically-oriented world).

So, all things considered, the initial question "What does it take to perform collaborative interdisciplinarity with good epistemic and academic results?" should be rephrased as "How can researchers who participate to an interdisciplinary research both achieve collective and interdisciplinary knowledge, in terms of establishing a hybrid methodology and developing general understanding of the broad project's results, and improve their disciplinary specialization, in terms of maintaining their disciplinary research updated and a high level of productivity?" In order to answer it, I believe that we should examine the academic and scientific environment in which an interdisciplinary collaboration might occur from a cognitive perspective that can make sense of dynamics of group cognition, of ecological and cognitive developments of the environment in which the collaboration happens, and of how cognitive activities are extended and distributed across artifacts, people, and spaces.

To address all these issues, I propose to adopt a broadly conceived distributed cognition theory to pragmatically address organizational and developmental problems of collaborative interdisciplinarity. This proposal comes for two reasons.

On the one hand the classical view of interdisciplinary as an exchange or an integration of knowledge from different disciplines is not enough focused on the human part of this phenomenon to give us operative solutions on how to plan an interdisciplinary research. Moreover, I will argue that seeing interdisciplinarity as merely an interaction between disciplines may mislead the necessary preparations for an interdisciplinary project. On the other hand, the studies on distributed cognition that so far have targeted interdisciplinary projects have been more concerned with the development and management of artifacts to pay attention to group co-ordination, and elements of ecological cognition.

So, by declaring that I will use a distributed cognition approach to address the typical and pragmatical issues of interdisciplinary, I need to specify that I will not concentrate my attention on the analysis of the cognitive role of highly developed artifacts. On the contrary, I will rather adopt an agent-based perspective which is focused on individual-to-group relationships and group cognitive issues. In particular I will focus on the embodied, emerging, and extended nature of cognitive activities at the core of successful examples of collaborative interdisciplinarity. To do that, I will employ particular concepts that emerged within or in relation to a broad description of distributed cognition, which incorporates insights from the extended mind approach, Gibson's ecological theory, and some hints of emerging cognition.

3 Why Do We Need to Face Interdisciplinarity from a Distributed Cognition Perspective?

To appreciate the refreshing perspective that distributed cognition and near views can bring to the analysis of interdisciplinary research, it may be useful to remind us what distributed cognition is and what it aims at. Rogers (1997, p. 1) gives us the hint in a well-condensed summary:

> Distributed Cognition is a hybrid approach to studying all aspects of cognition, from a cognitive, social and organizational perspective. The most well known level of analysis is to account for complex socially distributed cognitive activities, of which a diversity of technological artifacts and other tools and representations are an indispensable part.

The last part of this quote, which refers to *complex socially distributed cognitive activities*, could hardly be more precise in describing the cognitive processes in which collaborative interdisciplinarity is conceived, attempted, and sometimes achieved. Indeed, as already suggested, the analysis of scientific contexts and even interdisciplinary collaborations with the exploitation of insights from the distributed cognition framework has already been attempted and done. The literature on distributed cognition is full of references to the scientific framework and it also includes analyses of interdisciplinary work. The aim of these previous studies, though, was not focused on finding problem-solving strategies for issues related to interdisciplinary collaborations: the goal of various studies was to provide ways to look at the scientific and interdisciplinary environment as a stage for distributed cognition (as titles like "Scientific Cognition as Distributed Cognition" (Giere 2002), "Can Scientific Cognition Be Distributed?" (Rhee 2017) nicely indicate—cf. also Derry et al. (1998), Brown (2009)). In other words, it was a theoretical attempt to prove the foundation of the distributed cognition theory, not a way to reframe the practice of interdisciplinarity with suggestions from the theory of distributed cognition.

Nevertheless, this focus of the literature on distributed cognition mainly depended on the fact that distributed cognition framework, as well as near views that account for the emerging, extended and embodied features of cognition, are relatively recent perspectives in philosophy. But I agree with Rhee (2017) that even if there are still debates on whether cognition can be regarded as distributed or not, now it is natural to talk about distributed cognition in science and it would be more helpful to accept this hypothesis than taking more time to defend it. Indeed, I believe it could be useful from a pragmatical perspective, specifically it could help addressing problematic issues that often emerge in scientific contexts. So, instead of proving that interdisciplinary science is a framework where distributed cognition plays a huge role (or even that scientific cognition can be distributed), I will consider it as a given assumption, and I will exploit the literature on distributed cognition to focus on the cognitive activities in interdisciplinary collaborations as processes acting upon the cognitive exchange among researchers, and their interaction with the environment and technological devices.

Furthermore, I believe that considering the framework of distributed cognition at least appropriate to discuss issues emerging in interdisciplinary collaborations, could help understand the framework of interdisciplinary research as one that cannot be reduced to an exchange of *knowledge-that* (as already pointed out by Thagard (2006), exploiting the distinction provided by Ryle (1949)). Indeed, distributed cognition can rather focus on how the collaborative work involves many layer of trade of *knowledge-how*, procedural and declarative knowledge (respectively, specifiable rules and conceptual tools and schemes, (Anderson 1983)), and explicit and implicit knowledge. Moreover, at the same time I will also analyze socially-distributed cognitive phenomenons like organizational learning and the distribution of labor. Group cognition will also be taken into consideration, since, as affirmed already by Stahl (2010), even the first notes of the framework of distributed cognition were related to this topic:

> Cognitive processes may be distributed across the members of a social group. (Hollan et al. 2000, p. 176)

> The cognitive properties of groups are produced by interaction between structures internal to individuals and structures external to individuals. (Hutchins 1995, p. 262)

> The group performing the cognitive task may have cognitive properties that differ from the cognitive properties of any individual. (Hutchins 1995, p. 176)

Plus, I argue that by explicitly referring to the distributed cognition paradigm in order to address problems of interdisciplinary collaborations, a methodology could be adopted when a new interdisciplinary framework needs to be constructed: i.e. the more we will talk about interdisciplinarity as an object that would profit from an analysis based on the distributed cognition paradigm, the more people involved in interdisciplinary collaboration will be motivated in explicitly performing distributed cognitive activities. In turn, the more they will become aware and self-rewarded by the modification of their eco-cognitive environment.

Another argument in favor of a more self-aware interaction with the cognitive environment is provided by Thagard (2006). He argues that collaborations in scientific and academic projects require a considerable amount of procedural knowledge regarding how to collaborate, which can be described as the collection of know-how that allows the members of an interdisciplinary group to act to the benefit of the collaboration. This knowledge cannot be achieved just by analyzing abstract accounts of interdisciplinarity but it needs to emerge from explicit instructions regarding how relationships are built, on the distribution of knowledge between people, implicit and embodied role-modeling and, of course, in the emerging practices of functional teamwork. So, that is the reason why one of the sub-goals of this paper is to analyze and discuss those activities that can foster procedural knowledge about interdisciplinary collaborations as cognitive distributed systems and that may allow the individuals who need and want to collaborate to achieve more procedural knowledge during the project.

To exploit insights from distributed cognition to these aims, what I need to establish now is a way to look at issues that emerge in interdisciplinary collaborations that could make sense of them. Since the problems of collaborative interdisciplinarity are recorded since the 1970s, it is not so difficult to find the recurrent issues; moreover, a pattern that seems to give reasons for most of them is already comprehended by the consideration, still standing in academia, that interdisciplinarity is an exceptional plan for researchers. Indeed, a problem that often emerges in the literature on interdisciplinary collaborations refers to the fact that, even if interdisciplinarity is verbally encouraged since post-graduation by funding agencies and grant applications (Maki and MacLeod (2016) even call it a "dominant research policy imperative" given it prominent role in the statements of different funding agencies), few interactions between researchers belonging to different disciplines are not well-established enough to give directions to people new to the interdisciplinary framework.

So, while interdisciplinary collaborations are fairly common, the researchers that aims at planning one interdisciplinary project still find themselves in a new situation, without directions nor instructions to make it work. Moreover, since academic training is disciplinary-oriented, researchers are disciplinary-bounded to cultivate partial ignorance, which impedes researchers to understand and bridge the disparity between theirs and others' disciplinary perspectives. In the next section I will argue that this type of ignorance can be recognized in the common issues that emerge in interdisciplinary collaborations. So, exploiting insights provided by distributed cognition theories, we can design the cognitive system of interdisciplinary collaborations to oppose implicit, embodied, and tacit forms of disciplinary-bounded ignorance.

4 Disciplinary-Bounded Ignorance: Issues of Collaborative Interdisciplinarity

It is quite obvious that expectations drive most of our cognitive activities on a daily basis. We anticipate, plan, and expect, then we act on the bases of those implicit or explicit cognitive acts. We embodied our expectations, we modify our environment in order to distribute knowledge that we believe will turn out to be useful for our purposes, and, when hierarchies are well-established, team roles emerge from expectations of efficiency. Of course, expectations need to be adjusted in case we misunderstand or misconstruct a situation, making us lose time in the process. My claim is that, since expectations of researchers at the first experience of interdisciplinary work are disciplinary-bounded, they need to be repeatedly adjusted (and not always well adjusted) in the making of the collaboration. In turn, interdisciplinary collaborations become challenging, very time-consuming, and at first low rewarding for the researchers involved.

Indeed, expectations are built over past experience and research: unfortunately, past experience for a researcher does not always involve interdisciplinary work and when it does—and miraculously it is successful, it does not automatically give him/her instructions to plan another interdisciplinary research. Indeed, the intuitive approach to interdisciplinary research (which comes also from some abstract theories on interdisciplinarity) depicts it as a puzzle game, where a complex problem leads the way and everyone has a piece of knowledge that effortless contributes to solve it and to understand the big picture.

Moreover, regarding past experience, since academic training is disciplinary oriented we all are well-trained to see our discipline as the one that has ways to get answers for specific problems, but we are not used to see our disciplinary perspectives as a limited or partial view on the situation. This, as already mentioned, involves a form of ignorance that is bounded to the disciplinary framework of researchers, which shapes their expectations, and in turn increments the skepticism and lack of communication between them. Intuitively, bringing awareness to this form of ignorance and reducing it will ease the achievement of good epistemic results in interdisciplinary collaborations.

In order to reduce this form of ignorance, researchers should adopt strategies to share their epistemic background and express freely their epistemic and academic goals. These directions rely on the distributed nature of science, which can be spurred in particular and difficult situations—as the ones that usually emerge at least at the beginning of interdisciplinary collaborative research.

Evidence of disciplinary-bounded ignorance that permeates the expectations of researchers new to interdisciplinary work can be found in various reports of interdisciplinary experience, to which I will amply refer in the next subsections. One report is particularly valuable: a case study authored by MacLeod and Nersessian (2014, 2016), which extends over 5 years of investigation in two Integrative System Biology laboratories, located in a major university research institution.

Since I will extensively rely on the testimony of MacLeod and Nersessian to draw my case on disciplinary-bounded ignorance and distributed cognition strategies, I need to specify some details on their research first. They mainly investigate the cognitive richness of the interdisciplinary field of system biology, especially considering model-building strategies, cognitive-niche-curation actions, and the role of distributed cognition in the development of the research. Moreover, they also report and study the problem-solving strategies that researchers have adopted in the development of the interdisciplinary collaboration.

The tight collaboration between scientist in the Integrative System Biology laboratories was needed because integrative system biology aims at developing large scale mathematical models to map the structure and dynamics of complex biological systems. Hence, these researchers work two different but highly related problems: strictly mathematical tasks, such as developing new parameter-fixing algorithms for a given class of data and constructing models from data molecular biologists provide; and the set and run of experiments mean to establish the role of bio-molecular processes and elements found in the cellular environment. Thus, the work in the center relies on the collaboration between two classes of researchers: experimenters, who are molecular biologists with no training in modeling, and modelers, coming mostly from engineering fields, who confessed to never have had experimental training.

Now, it is fairly intuitive to say, and it is confirmed by the reports of MacLeod and Nersessian (as well as others[3]), that all researchers who approach interdisciplinary work have expectations regarding three main types of knowledge. First: the knowledge that they brought to the project, in terms of specialized expertise: for example which kind of routines they expect to perform, which technological instruments and data they will need, and how they will present their works to colleagues. Second: the knowledge of their epistemic goals, which encompasses understanding the main goal of the project as well as their smaller academic goals, such as the amount of publications they want/need to bring home from the experience, and what kind of

[3]For reports on the expectations of researchers at the beginning of interdisciplinary work see: Buller (2009), Weingart (2000), Andersen (2013), Donaldson et al. (2010), Fiore et al. (2008), Rose et al. (2011), Lopez (2015), Grune-Yanoff (2016).

results are expected month-to-month or a-year from the research. Third, of course, they assume also that they will learn from the interdisciplinary experience, thus they will acquired new methodologies and that would improve their roles and proficiency in future projects and programs.

Many of the problems that emerge at the beginning of interdisciplinary collaborations depend on the disparity between the expectations regarding those three type of knowledge researchers have and what they actually need, have, or get during the collaboration. A couple of quotes, taken from the reports of MacLeod and Nersessian (2014, p. 233), summarize well these mismatch between expectation of researchers and the actual situation, which fueled cross miscommunications and impasses in the laboratory. The first came from a modeler who was modeling atherosclerotic inflammation:

> the biologist(s) produce the data they want. But those data are not actually what we want when we do parameter estimation... so there... might be some gap between these two, between us... they only focus on one species. But even so they don't produce enough data.... they don't measure the concentration for example. And they have few kinetic data...

The second complain came from an experimental collaborator:

> the data that they [modelers] want from us is something that is not simple to generate. So if they want a Km for an enzyme we have to purify the enzyme. Then we have to create all the conditions to measure it in vitro. That's not a simple undertaking. That's probably six months of work. And none of us have a person sitting around who can do that for six months. So that's the first problem. The second problem is yeah if we are going to do that. If we are going to spend six months generating what they want then we would like we need to have something that is going to come out of it.

These two quotes represent three different levels of problems that recur in reports of interdisciplinary collaboration: (1) putting together researchers from different disciplines means putting together specialists that have different methodologies, practices, and values; (2) since disciplinary background have always determine the timeline and reasons for establishing the epistemic goal, it is difficult that researchers find there a common ground; (3) since background knowledge and epistemic/academic goals seem to differ, communications, as well as learning, become harder and harder the more impasses and difficulties emerge. These three level are embodied in the researchers practices, are distributed across objects, other people, and the environment, and emerge as particular issues in the interaction between members of the same research group. Indeed, as presented by Giere:

> what philosophers typically regard as 'background knowledge' or 'auxiliary assumptions' may not best be understood in terms of symbolic representations but as being physically encapsulated in experimental apparatus. [...] it is particularly enlightening to think of the whole facility as one big cognitive system comprised, in part, of lots of smaller cognitive systems. [...] A better description of the situation [which refers to a laboratory in the Indiana University Cyclotron Facility] would be to say that the data is being gathered by a complex cognitive system consisting of the accelerator, detectors, computers, and all the people actively working on the experiment. (Giere 2002, p. 6)

So, a way to look at the issues that emerge from the disparity between expectations of researchers and the reality of the interdisciplinary collaboration is to look at how researchers extend their knowledge into the environment, how they embody their expectations regarding academic and epistemic goal, and how they assume that collaboration would emerge as a trait of the project. Focusing on those performances, it would also be useful to consider how, sometimes, they try to react to impasses and miscommunication in order to make the interdisciplinary research work.

To this aim, in the next subsections I will comment the reports on the expectations of researchers new to interdisciplinary work, the actual amount of collaborative work that the interdisciplinary research needs, and the strategies that aimed at fixing the gap between expectations and reality. In the first subsection, I will examine the problem regarding what kind of knowledge every researcher puts into the interdisciplinary collaboration and why it needs to be rethought in a more locally-distributed and extended perspective. In the second subsection I will discuss how expectations regarding goal-oriented knowledge can be more usefully exploited if they foster new affordances for the entire group, and not just the disciplinary-divided researchers. Finally, in the third subsection I will discuss the problem of interdisciplinary learning, why expectations regarding it usually involve the problematic phenomenon of *the illusion of depth of understanding* (Ylikosky 2009), and how a distributed cognition point of view can help researchers establish a good interdisciplinary program without sacrificing specialized expertise.

4.1 Bridging Different Background Knowledge Through Co-location and Small Groups Composition

One of the key problems that affects the development of a collaborative interdisciplinary project regards the fact that the specialized expertise of the members of the team is, in a way, both highly valuable and an impediment to the good establishment of common and cross-disciplinary knowledge. MacLeod and Nersessian (2014, 2016) give a lot of examples of miscommunications and impasse between modelers and biologists that depended on the different background assumptions that they had on methodological steps and time schedules. Basically, they had specialized epistemic backgrounds that blocked them to fully understand the one belonging to the others and impede the others to understand theirs.

For instance, often modelers could not work with data biologists provided because they claimed they were not relevant for the model construction, while biologists kept complaining about the absurd requirements of modelers, that expected data that were actually very hard to get from the usual experiments performed. The frustration between the two groups grew as communications became difficult and the beginning of the collaboration was recorded as a highly stressed period.

They did basically need to achieve what Collins and Evans (2002) call “interactional expertise”. Interactional expertise describes the competence which derives from knowing enough about a field's methodology to be able to engage with it productively, without necessarily being able to operate within the field. Even if it would be incredibly useful in interdisciplinary studies, according to Collins and Evans the mutual understanding that comes from it is attained through a long process of socialization and sharing that can be hardly produced fast enough in the daily practice of an interdisciplinary research. Thus, using this idea as a conceptual compass, I've analyzed those strategies that have led to functionally organize and improve the epistemic productivity and the relationships between researchers in others case studies.

The ground assumption is that in an interdisciplinary project the epistemic backgrounds of researchers differ, as they lack shared methods, practice, values, which comprehend the terminology to express the missing pieces. Thus, they need ways to implement not mere communication, but a common epistemic and ecological ground, from which they could build trust and understanding and the establishment of fair relationships of epistemic dependency (Hardwig 1991). So, in order to determine a fair relation between specialists from different disciplinary background, a few insights can be exploited from distributed cognition approach.

A first consideration that can be useful to discuss here is the assumption (discussed by Giere (2002) to address features of group cognition in scientific fields), that cognitive systems consisting of a group of individual possess knowledge that is both highly valuable and redundant. In an interdisciplinary collaboration, finding the redundant knowledge is both needed to get the project started and to assemble a team with a common vocabulary and common knowledge. Helpful in these cases are boundary objects, defined by Star and Greisemer (1989), Donaldson et al. (2010) as "something that maintains its identity across a range of different communities of practice, providing a commonality that allows them to work together." Also nondisciplinary objects are precious, which are involved in all organizational and coordinating activities. According to Donaldson et al. (2010) both boundary and nondisciplinary objects stand out from the "messy" objects, which are disciplinary specific and cause incomprehension in the interdisciplinary research.

So redundant knowledge can obviously regard what Thagard (2006) calls the procedural knowledge of researchers with reference to collaborations. Procedural knowledge is the tacit set of skills and knowledge that researchers get and invest during their work routine. Basically, even if those researchers never have worked with peers and superiors from other disciplines, they probably have experience in collaborative work with specialists from their own field. That kind of collaborations necessarily have some common features to the one that they are trying to achieve in the interdisciplinary project. Explicitly calling for those features, for instance by organizing meeting with the clear intent of discussing experience in cooperative learning and collaborative work, is reported to have a good impact on the establishment of a team. By using this strategy, the procedural knowledge on collaboration can actually be distributed across different specialists and a negotiation of expertise can be traded in the group network.

Moreover, it could be useful to address communication problems of interdisciplinarity groups as problems of knowledge poorly distributed. For example, co-location is an incentive to improve communication and learning: as suggested by

Stahl (2010), constructing a common environment in which all groups of expert would benefit in addressing disciplinary and interdisciplinary problems could help them develop clearer path of communication. As Boden (1997, p. 18) highlighted:

> Physical co-location is very helpful. However, site-visits and Internet-grouping can compensate for geographical distance to some extent. Frequent meetings between the different disciplines in the teams are essential. Good informal communications are valuable also. [...] Co-location of professionals in different fields is likely to encourage open-mindedness in the students, even if their research does not call on all the disciplines.

For the development of an interdisciplinary group, also Derry proposes to considering the sharing of cognitive space as a way to foster communication and openness between researchers:

> To illustrate the shaping forces of such contexts, consider the typical case of a working group whose members have offices in different buildings. In this context the group might come to rely heavily on e-mail and periodic formal meetings for communication. Situated cognition would predict that this same group would develop significantly different practices if its members all worked in physical closeness to one another. For example, they might develop strong social ties and employ only informal, face-to-face communication. (Derry et al. 1998, p. 27)

Moreover, reconfiguring the ecological space of the interdisciplinary collaboration and the small groups that need to work together means also reconfiguring the cognitive division of labor, which supports a hierarchical distribution of resources (Giere 2002, p. 1). More than that, a good communication at the beginning of the group foundation can be favored by using theories of group cognition, as the rich knowledge-sharing properties of small groups. Indeed, as argued by Stahl (2006, p. 16):

> Small groups are the engines of knowledge building. The knowing that groups build up in manifold forms is what becomes internalized by their members as individual learning and externalized in their communities as certifiable knowledge.

So, supporting the formation of small groups (which, according to Stahl, amount at groups larger than dyads included in the larger community) can be a good strategy to predispose activities of collaborative learning. The size of the groups matters since the individuals are more encouraged to share needs and to propose courses of actions in a smaller party, and then proceed to address the large group as a team. In fact, the small size group can prospect more achievable goals than the community at large.

Following the alterations of space and group dynamics in Macleod and Nersessian's study, they indeed speak of a division between older researchers that were teachers and repositories of knowledge, and younger researchers, with less specialization, who became more useful in the interdisciplinary team. These young researchers became the preferred path of communication between the two categories of researchers (and the smaller groups that emerge among them). They helped constructing what Zollman (2007) calls a "specialized knowledge network," where everyone receives all the information the group at large need, but through some differently competent people.

The specialized knowledge network, in turn, fosters regular and informal communication, which is recommended by different authors that succeeded in assembling an interdisciplinary team and make it work for the long run (Boden 1997; Naiman 1999; Thagard 2006; Paletz et al. 2016).

At last, another phenomenon that is highly related to a good exchange of informal communications and relationships (especially at the beginning of a collaboration) is what Thagard (2006, p. 191) calls emotional contagion. This, which is defined as what "occurs when people automatically mimic and synchronize their movements with the facial expressions, voices, postures, movements, and instrumental behavior of others" can be seen both as the result of a good team assembling and as the indicator of a potential (but not yet established) fruitful collaboration. In turn, Thagard even argues that fostering emotional contagion is part of the procedural knowledge that members of an interdisciplinary group can learn by adopting conscious choices related on the amount of emotional engagement they mean to distribute into the group environment.

So, addressing the difference in background knowledge can be fostered by choosing co-location of the groups and the establishment of small parties in the large team. With these tactics, tacit and explicit knowledge can be more easily distributed (so integrated) in the the interdisciplinary research as a developing eco-cognitive system. Of course, sharing background knowledge means also recognizing different but not necessarily mutually exclusive goals. To address the difference between dissimilar and potentially contrasting goals, encouraging strategies can be exploited by considering the theory of *affordances* (Gibson 1979) and by avoiding what Werner (2019) calls *cognitive confinement*.

4.2 Goals Projections and the Problem of Cognitive Confinement

Another frequent issue that is reported in the literature on interdisciplinary collaboration is the fact that results take an unexpected longer time to be published, since the interaction on the group requires adjusting to other's methodologies, vocabulary, and literature. Moreover, as variously reported (Naiman 1999; Andersen 2013; Paletz et al. 2016) the difference in intellectual caliber and commitment to team-oriented research of the members of the interdisciplinary group heavily affects the research in terms of timeline projection, task-oriented sub-projects and team building. This problem usually emerges because the leader of the team must bear more of the load of the group and needs to make difficult choices during the entirety of the project, especially regarding the composition of the group (and smaller sub-groups). All these issues relate to the problem that concerns the disciplinary-oriented approach that most researchers have to their epistemic and academic goals (Donaldson et al. 2010). In the case commented by MacLeod and Nersessian, they explicitly recognize this issue by saying:

> Molecular biologists have their own disciplinary aims and agendas, which don't necessary align with the requirements of modeling. (MacLeod and Nersessian 2016, p. 410)

In that case the individual aims collided, for example, when some biologists required a specific model for a particular set of data in order to publish a paper on them, and modelers needed larger amount of data in order to publish a paper on the application of that particular model for different variables. Even if the creation of that model was necessary for both the categories, the unexpressed small epistemic goals generated longer timetables for results and even altered the main project timelines. The impasses arose so often and the communication was so difficult that researchers felt isolated and kept questioning their role in the project. Now, even if this seems an extreme situation of bad communication strategies, these reports describe patterns of malfunctioning interaction that can be found in all the literature over collaborative interdisciplinary.

A way to look at these dysfunctional communication is by referring to the phenomenon of cognitive confinement, which is defined by Werner (2019) as which is "a state of being structurally or systematically unable to gain information from an environment, determined by patterns of interaction between the subject and the world." Information, as Werner uses it, is a broad notion that incorporates takes from distributed cognition theory and the famous notion of affordances coined by Gibson (1979), which, for my limited purposes here, can be described as "opportunities for actions" that can be immediately perceived by an agent into objects, other people and environmental traits. A phenomenon of cognitive confinement is so described as a systematical impediment to get specific affordances in an environment that could offer them if the epistemic or behavioral dispositions of the agent were different.

To offer a concrete example I can refer to the situation, very common in the interdisciplinary literature, in which researchers do not talk about their personal academical or research goals (number and type or papers that they want to publish, experiments that they need to do, research outside the field of the interdisciplinary research they want to conduct in the meantime) to the specialists outside their discipline during the project tasks projection. The commonly understood cause of this lack of communication is the default assumption that, if a particular goal is not already part of the interdisciplinary project, it would take time or resources from it, and researchers think that it would be best for them to aim at it in their personal time. The phenomenon of cognitive confinement can easily explain the difficulty of those researchers in obtaining that individual academic or research goal because, by omitting to discuss them with the group, they limit their options in terms of time and resources spent in their research, they minimize possibility of interaction between them and others, and they refrain from discussing availability problems with the interdisciplinary group. In the case studied by MacLeod and Nersessian, a contributing factor to this situation was that, at the beginning of the project, the aim of the laboratory was just discussed by the directors, without consulting the permanent or temporary collaborators, who had all the interest in getting more published paper done by the end of the collaboration.

A solution to these problems emerged when the discussion of individual goals became an integral part of the research meeting. Indeed, a pattern of collective behavior began to emerge and molecular biologists recognized that they could take some time from the interdisciplinary research to publish more. They also developed what MacLeod and Nersessian (2016) called "adapting problem solving" strategies to complete them without affecting the development of the project. Doing that, they reframed the cognitive system of the interdisciplinary research in order to get more chances and affordances that were previously locked for lack of goal-oriented communications. This improved not only the individual and specialized knowledge of researchers, but it even encouraged disciplinary-explicit behavior in the interdisciplinary environments of the two laboratories.

4.3 Cross-Disciplinary Learning and the Illusion of Depth of Understanding

At last, the ultimate goal for researchers who embark into an interdisciplinary project is to learn: to learn new ways to approach particular issues, to learn ways to reconfigure their research study for outsiders of one's field, and, in best case scenarios, to learn how to plan another interdisciplinary research or to find a position in a different one. Luckily, one common trait of all interdisciplinary works—successful and less so—seem to be the cross learning that happens between researchers. In worst cases, at least researchers learn to recognize the symptoms of a bad collaboration or of a stalled interdisciplinary project. Needless to say, though, when discussing learning, the expected and the actual amount of effort that someone needs to put in the process vary greatly on the ground of personal character, career progress, and power position in the group. Moreover, seemingly no-one is fully prepare to the actual work that understanding others' disciplinary methodology involves, even if they started the research with this conscious target in mind. As Thagard (2006, p. 185) records:

> Probably the most important piece of relevant declarative knowledge is that people from different disciplines may vary a great deal in their methods and conceptual schemes, so that the same terms may have different meanings. [...] The procedural lesson is that if you want to do interdisciplinary, peer-different research, then you need to be willing to spend a great deal of time listening to and learning from your collaborators in order to find out how they think about things. Stating this as a rule does not fully capture the acquisition of the kind of patient, attentive, conversational style that is necessary to elicit mutual understanding.

Indeed, many authors[4] report that one of the main obstacles that emerges at the beginning of the interdisciplinary collaboration is the disparity between the amount of time that researchers expected to spend in learning about other disciplines and their vocabulary, and the necessary time they needed to spend in order to become actually functional in interdisciplinary interactions. Moreover, many researchers complained

[4]Just to name a few of them: Naiman (1999), Thagard (2006), Andersen (2013), Paletz et al. (2016).

that their own specialization suffered from the time spent trying to understand other disciplines' requirements (Naiman 1999).

The disparity between expectations and reality regarding one's own learning process is affected not only by the absence of a consistent literature on the do's and don'ts on how to plan an effective interdisciplinary research as a rich distributed system, but also by cognitive biases that affects the way researcher distribute knowledge into their environment. One that can impede a productive and honest communication between researchers involved into an interdisciplinary research is the Illusion of Depth of Understanding (studied with reference to the scientific context by Ylikosky (2009)). This phenomenon describes people's tendency to overestimate the detail, coherence, and depth of their understanding. The name of this effect has been coined by Keil (2003), who, with some colleagues (Rozenblit and Keil 2002; Mills and Keil 2004) experimentally investigated the influence of this effect and found that most people are prone to feel that they understand the world with greater detail, coherence, and depth than they actually do. The explanation to this tendency is found in the fallibility of the metacognitive ability that inform us of having understood something: the *sense of understanding*. The sense of understanding is a special kind of feeling of knowing which is highly fallible (Grimm 2009) and can be easily confused with what it should indicate: actual understanding. Indeed, understanding and the sense of understanding are only flimsily related: sometimes the sense of understanding can lead the agent to overestimate her understanding of something and sometime it can lead to underestimate it. Usually the former is the case, but not always.

The Illusion of Depth of Understanding is important if discussed in relation to the problematic communications between researchers because the sense of understanding occurs primarily for explanations if compared with facts, procedures, and narrative (Rozenblit and Keil 2002; Keil 2003). Understanding how to act into or plan ahead in a collaborative project depends on the explanations that different specialists need to give to each others, especially explanation given to and by people outside their own field. Those explanations can lead to a poor—and sometime inexplicably poor—judgment on how an outsider of one's discipline understood what others need to do next and how it will affect their work plan and schedule.

The sense of understanding plays a similar role to the feeling of knowing: it should inform the agent about when she has enough data to say that she understand something. Usually, it leads the agent to overestimate her understanding because in ordinary circumstances having a comprehensive understanding of something is impractical: having first the sense of understanding misleads the agent to think about understanding as an on-and-off phenomenon instead of one that comes in degrees (Ylikosky 2009) and when one reassures herself that she understands something, she can act about it. Basically, quicker the explanatory reasoning kicks-in the sense of understanding, quicker the agent feels that she needs no more data, and she can act about what she thinks she understood.

A way to partially explain the miscommunications and impasses that block—especially at the beginning—the development of a productive interdisciplinary project is by considering the hypothesis that the brief explanations that different specialists give to other researchers outside their field is not enough to instill actual understanding but just the most similar version of it in their minds: the illusion of depth of understanding. Actual understanding is a condition that involves different stages (Ylikosky 2009) and some authors argue that it necessitates having an internal mental model of the object of understanding (Waskan 2006). Since an interdisciplinary project do not automatically provide the researchers the conceptual tools to build an internal model for the epistemic, technical, and methodological requirements of all the disciplines involved, it is clear that they need to compensate with short cut their cognitive limitations and initiate the project half-understanding what others need or do. This illusion of course, not only affect their interaction with others specialists in the project, but it also prevent a useful distribution of more information and knowledge regarding disciplinary-bounded objects, vocabulary, goals, and timeline in the general project.

One strategy that can limit or at least help to address this problem is to plan few weeks-long training events for the whole team before the beginning of the project (as suggested by Donaldson et al. (2010), Chandrasekharan and Nersessian (2015)). This action, in particular, rewarded the researchers involved in the case followed by MacLeod and Nersessian. They built an introductory modeling course for bioscientists, conducted by the modelers of the laboratory. The course helped the modelers to reframe their background knowledge in order to explain its basis to non-specialists, and the biologists to understand the requirements of model-building, in order to develop experiments that could fit some of its basic criteria. At last, thanks to this course, they developed a relationship of trustful epistemic dependence: biologists stopped relying by pure faith on the competence of modelers and found comfort in having a little degree of control on what modelers were doing with their data.

Thus, the course provided a way to reframe the knowledge of modelers for the colleagues, implementing useful interactional expertise in the team members and fastening the distribution of collective and interdisciplinary knowledge in the interdisciplinary eco-cognitive system.

5 Conclusions

So, "what does it take to perform collaborative interdisciplinarity with good epistemic and academic results?" The strategies that different groups can adopt to get an interdisciplinary research work, obviously, cannot all be gathered and be thoroughly explained in a small paper. However, this brief article was not aiming at giving an exhaustive list of instruction to apply in order to get an interdisciplinary research going and in the right direction. Instead, I wanted to get a new perspective to address this pragmatical question and get new answers from recorded experiences of interdisciplinary collaborations.

Exploiting the perspective of distributed cognition means focusing on the cognitive system that an interdisciplinary project develops, and looking at the recurrent issues as pattern of cognitive failures that relates to features of group cognition, and extended and distributed cognitive activities. The strategies proposed follow the assumption that the cognitive system that ground an interdisciplinary research can be improve by addressing and possibly avoiding cognitive biases and behaviors that limit the necessary distribution of knowledge across other researchers and the extension and integration of that knowledge into the common environment. Nevertheless, since this paper is a theoretical analysis of reported cases of interdisciplinary collaborations, I believe this work should be followed an empirical testing of the hypotheses here presented.

There are obviously other issues that can affect an interdisciplinary collaboration that can be discussed adopting a distributed cognition approach. For example, I did not address the psychological and sociological problems that that depend on personal issues and the interpersonal skills of the members of the interdisciplinary team. As noted by Naiman (1999, p. 293): "a lack of patience, understanding, trust and respect, or unspoken jealousies and perceived threats to someone's position or authority can be devastating." However, in this particular framework I could not address these problematic traits without an alarming degree of generalization. The reports of these kind of problems are too few to be considered in a pattern-establishing study and the fact that personal feelings and interpersonal skills are underestimated in the philosophical and sociological literature does not help in finding good ways to approach these issue. However, bridging this gap could also be a useful follow-up to this research.

References

Ackoff RL (1974a) Redesigning the future. Wiley, New York

Ackoff RL (1974b) The systems revolution. Long Range Plan 7:2–5

Andersen H (2013) The second essential tension: on tradition and innovation in interdisciplinary research. Topoi 32:3–8

Anderson JR (1983) The architecture of cognition. Harvard University Press, Cambridge

Apostel L, Berger G, Briggs A, Michaud G (eds) (1972) Interdisciplinarity: problems of teaching and research in universities. Organization for Economic Cooperation and Development, Paris

Boden M (1997) What is interdisciplinarity? In: Cunningham R (ed) Interdisciplinarity and the organisation of knowledge in Europe. Office for Official Publications of the European Communities, Luxembourg, pp 13–26

Bromham L, Dinnage R, Hua X (2016) Interdisciplinary research has consistently lower funding success. Nature 534:684–687

Brown MJ (2009) Science as socially distributed cognition: bridging philosophy and sociology of science. In: FrançSois K, Löwe B, Müller T, Van Kerkhove B (eds) Bringing together philosophy and sociology of science. Foundations of the Formal Sciences VII, New York, pp 285–299

Buller H (2009) The lively process of interdisciplinarity. Area 41(1):395–403

Chandrasekharan S, Nersessian NJ (2015) Building cognition: the construction of computational representations for scientific discovery. Cognit Sci 39(8):1727–1763

Collins H, Evans R (2002) The third wave of science studies studies of expertise and experience. Soc Stud Sci 32:235–296

Derry SJ, DuRussel LA, O'Donnell AM (1998) Individual and distributed cognitions in interdisciplinary teamwork: a developing case study and emerging theory. Educ Psychol Rev 10:25–56

Donaldson A, Ward N, Bradley S (2010) Mess among disciplines: interdisciplinarity in environmental research. Environ Plan 42(A):1521–1536

Fiore SM, Hoffman RR, Salas E (2008) Learning and performance across disciplines: an epilogue for moving multidisciplinary research toward an interdisciplinary science of expertise. Mil Psychol 20(1):155–170
Gibson JJ (1979) The ecological approach to visual perception. Houghton Mifflin, Boston, MA
Giere P (2002) Scientific cognition as distributed cognition. In: Carruthers P, Stich S, Siegal M (eds) The cognitive basis of science. Cambridge University Press, New York, pp 285–299
Grimm SR (2009) Reliability and the sense of understanding. In: Regt HD, Leonelli S, Eigner K (eds) Scientific understanding: philosophical perspectives. University of Pittsburg Press, Pittsburg, pp 83–99
Grune-Yanoff T (2016) Interdisciplinary success without integration. Eur J Philos Sci 6:343–360
Hardwig J (1991) The role of trust in knowledge. J Philos 88(12):693–708
Holbrook JB (2013) What is interdisciplinary communication? Reflections on the very idea of disciplinary integration. Synthese 190:1865–1879
Hollan J, Hutchins E, Kirsh D (2000) Distributed cognition: toward a new foundation of human-computer interaction research. ACM Trans Comput-Hum Interact 7(2):174–196
Hutchins E (1995) Cognition in the wild. The MIT Press, Cambridge, MA
Jantsch E (1970) Inter- and transdisciplinary university: a systems approach to education and innovation. High Educ 1(1):7–37
Keil FC (2003) Folkscience: coarse interpretations of a complex reality. Trends Cognit Sci 7:368–373
Klein JT (1990) Interdisciplinarity: history, theory and practice. Wayne State University, Detroit
Koskinen I, Maki U (2016) Extra-academic transdisciplinarity and scientific pluralism: what might they learn from one another? Eur J Philos Sci 6:419–444
Lachance C, Larivière V (2014) On the citation lifecycle of papers with delayed recognition. J Inform 8(4):863–872
Larivière V, Haustein S, Börner K (2015) Long-distance interdisciplinarity leads to higher scientific impact. PloS ONE 10(3):e0122565
Lopez WL (2015) Interdisciplinariety: a perspective from the dynamics of scientific production and communication. Univ Psychol 14:1–2
MacLeod M, Nersessian NJ (2014) Strategies for coordinating experimentation and modeling in integrative system biology. J Exp Zool 322:230–239
MacLeod M, Nersessian NJ (2016) Interdisciplinary problem-solving: emerging modes in integrative system biology. Eur J Philos Sci 6:401–418
Maki U (2016) Philosophy of interdisciplinarity. What? Why? How? Eur J Philos Sci 6:327–342
Maki U, MacLeod M (2016) Interdisciplinarity in action: philosophy of science perspectives. Eur J Philos Sci 6:323–326
Merton RK (1968) The Matthew effect in science. Science 159:56–63
Metzger N, Zare RN (1999) Interdisciplinary research: from belief to reality. Science 283:642–643
Mills CM, Keil FC (2004) Knowing the limits of one's understanding: the development of an awareness of an illusion of explanatory depth. J Exp Child Psychol 87:1–32
Naiman R (1999) Interdisciplinarity in philosophy of science. Ecosystems 2:292–295
Naiman RJ, Bisson PA, Lee RG, Turner MG (1998) Watershed management. In: Naiman RJ, Bilby RE (eds) River ecology and management: lessons from the Pacific coastal ecoregion. Springer, New York, pp 642–661
Paletz SBF, Chan J, Schunn CD (2016) Uncovering uncertainty through disagreement. Appl Cognit Psychol 6:343–360
Rhee YE (2017) Can scientific cognition be distributed? Ann Jpn Assoc Philos Sci 26:29–37
Rogers Y (1997) A brief introduction to distributed cognition. http://www.slis.indiana.edu/faculty/yrogers/papers/dcog/dcog-brief-intro.pdf
Rose LT, Daley SG, Rose DH (2011) Let the questions be your guide: Mbe as interdisciplinary science. J Compil 5(4):81–91
Rozenblit L, Keil FC (2002) The misunderstood limit of folk science: an illusion of explanatory depth. Cognit Sci 26:521–562

Ryle G (1949) The concept of mind. Hutchinson, London
Stahl G (2006) Group cognition: computer support for building collaborative knowledge. MIT Press, Cambridge
Stahl G (2010) Group cognition as a foundation for the new science of learning. In: Khine MS, Saleh IM (eds) New science of learning. Springer, Canada, pp 23–44
Star SL, Greisemer JR (1989) Institutional ecology, 'translations' and boundary objects: amateurs and professionals in Berkeley's museum of vertebrate zoology. Soc Stud Sci 19(1):387–420
Strathern M (2004) Commons and borderlands: working papers on interdisciplinarity, accountability and the flow of knowledge. Sean Kingston Publishing, Wantage, Oxon
Thagard P (2006) How to collaborate: procedural knowledge in the cooperative development of science. South J Philos 44:177–196
Thorén H, Persson J (2013) The philosophy of interdisciplinarity: sustainability science and problem-feeding. J Gen Philos Sci 44:337–355
Waskan JA (2006) Models and cognition. The MIT Press, Cambridge, MA
Weingart P (2000) Interdisciplinarity: the paradoxical discourse. In: Weingart P, Stehr N (eds) Practicing interdisciplinarity. University of Toronto Press, Toronto, pp 25–42
Werner K (2019) Cognitive confinement: theoretical considerations on the construction of a cognitive niche, and on how it can go wrong. Synthese. Special issue knowing the unknown (page forthcoming)
Ylikosky P (2009) The illusion of depth of understanding in science. In: Regt HD, Leonelli S, Eigner K (eds) Scientific understanding: philosophical perspectives. University of Pittsburg Press, Pittsburg, pp 100–119
Zollman KJS (2007) The communication structure of epistemic communities. Philos Sci 74(1):574–587

Humor and Ignorance in the Perspective of Cognitive Niche Curation

Tommaso Bertolotti

Abstract The Incongruity Theory of humor suggests that humor is triggered by the perception of something incongruous. The recognition of ignorance amounts to recognizing an incongruity between one's actual knowledge (or lack thereof) and what ought to be known about that particular issue. In many cases, the explicit or implicit recognition of ignorance is cause for humor and laughter. In this paper, I will propose that humor can be understood as a social modality of signaling an instance of ignorance and urging its resolution: referring to the theory of cognitive niche construction, I will claim that humor out of ignorance is a niche curation technique, aimed at improving the quality of the niche by keeping the ignorant in line but without excluding them from participating to common niche-construction activities.

Keywords Ignorance · Applied epistemology · Cognitive niche construction · Humor

1 Introduction

Philosophers, educators, and broadly speaking most cultured people would agree that ignorance is no laughing matter. Ignorance *should not be* a cause for laughter. What is at stake is not the kind of ignorance that is clearly the elbow room for scientific progress or any other creative endeavor, so to say "positive" ignorance. We are talking about sheer ignorance: about not knowing (practically or theoretically) something that you really ought to know, that you should have known and you could have known. That kind of ignorance is often a laughing matter. The history of culture is plentiful with such occurrences: from Thales falling into a ditch while stargazing and being mocked by a charming and witty maid, to Andersen's naked emperor stirring up the laughter of his subjects. Quite curiously, we see that human being

T. Bertolotti (✉)
Department of Humanities, Philosophy Section, University of Pavia, Pavia, Italy
e-mail: bertolotti@unipv.it

Mind & Society Center, University of Southern California, Los Angeles, USA

T. Bertolotti (ed.), *Cognition in 3E: Emergent, Embodied, Extended*,
Studies in Applied Philosophy, Epistemology and Rational Ethics 56,
https://doi.org/10.1007/978-3-030-46339-7_2

have always valued health, wealth, and knowledge, but while sickness and poverty are not funny, ignorance can be hilarious.[1] Why is that so?

The aim of this paper is to answer this question by relying on cognitive niche construction theories (Bertolotti and Magnani 2017; Clark 2005), and on the incongruity theory of humor introduced by Aristotle and fine-tuned by Schopenhauer (Clark 1970).

2 Ignorance in Cognitive Niche Construction

2.1 Cognitive Niche Construction and Curation

Cognitive niche theories consist in a theoretical framework that has proven extremely profitable in bridging evolutionary biology, philosophy, cognitive science, and anthropology by offering an inter-disciplinary ground causing novel approaches and debates to crucial issues in all of the aforementioned fields (Clark 2005; Iriki and Taoka 2012; Magnani 2007, 2009; Odling-Smee et al. 2003; Pinker 2003; Tooby and DeVore 1987; Wallach 2015).

In spite of their differences, reviewed in Bertolotti and Magnani (2017), most cognitive niche theories maintain that one of the main characteristics of *Homo sapiens sapiens*'s development (and uniqueness) rests in their ability to use the environment in a way that is *cognitively relevant*. In other words, human success primarily relies on their ability to gather and exchange information and knowledge from the environment (as for cognitive niche theorists), and on their ability to alter the environment in order that it better serves cognitive scopes (in the view of cognitive niche *constructionists*). These modifications come in the form of knowledge distribution with little ecological relevance, capable of providing valuable scaffolding for the cognitive operations on which the success of human endeavors depends.

For the current scope, Clark's theory is the most relevant conversely, as he pays crucial attention to the local, ecological dimension of cognitive niche construction.

> "cognitive niche construction" [is] hereby defined as the process by which animals build physical structures that transform problem spaces in ways that aid (or sometimes impede) thinking and reasoning about some target domain or domains. These physical structures combine with appropriate culturally transmitted practices to enhance problem-solving, and (in the most dramatic cases) to make possible whole new forms of thought and reason. (Clark 2005, pp. 256–257)

[1] For the argument's sake, the kind of ignorance I am referring to is negative ignorance, i.e. "lack of knowledge". Still, the disanalogy between wealth and health, on the one side, and knowledge, on the other, holds also if we consider ignorance as a *blank* to be filled by learning or research: learning and researching are intrinsically exciting in their open-ended-ness. Healing to recover from a lack of health and getting rich to break free from poverty are less intrinsically pleasurable, as one would just like to attain the goal state as soon as possible, or not having to attain it at all.

Clark exemplifies this by telling of a busy bar where a bartender, upon receiving orders, arranges the glasses in front of her according to their shape, and adds decorations such as colored straws and cocktail umbrellas, in order to better remember the next drinks she has to mix without the support of a written list. It is not the human species' cognitive niche concerned here, it is the bartender's own niche that she is constructing by herself.

We modify local environments, for instance by using pencil and paper to draw a map, but we also develop and hang signs that are as universally understandable as possible, in order to make life easier cognition-wise. According to Clark, one of the first and most impressive instances of cognitive niche construction is the deployment of language into our ecologies, which establishes a powerful scaffolding for more complex cognitive endeavors.

Tooby and DeVore propose a theory of cognitive niches that is very relevant to the epistemological discourse (Tooby and DeVore 1987). According to them, the cognitive niche is about uncovering and exploiting, in a persistent way, cause-effect models of the external world in order to map successful behavioral possibilities.

> At the core of this lies a causal or instrumental intelligence: the ability to create and maintain cause-effect models of the world as guides for prejudging which courses of action will lead to which results. Because there are an infinitely large number of possible sequences of behavior (of which the overwhelming majority are maladaptive) "behavioral flexibility" is an insufficient characterization of our innovative adaptive pattern. Our cognitive system is knowledge or information driven, and its models filter potential responses so that newly generated behavioral sequences are appropriate to bring about the desired end. Of course, exploration, trial and error, and feedback are essential to the system, but, by themselves, they are inadequate to construct or maintain it. (Tooby and DeVore 1987, p. 210)

Cognitive niches must not only be constructed, but once in place they have to be maintained and updated. Let us consider road circulation as a cognitive niche: a YIELD sign is a cognitive distribution into the environment affording drivers some better decision making hopefully their fitness (i.e. not suffering a major car accident). Imagine that the circulation in a neighborhood is changed, so that the drivers who had to yield before now the right of way. It is then imperative not to leave in place the old YIELD sign, because whereas All-Way STOP intersections make sense, All-Way YIELD do not. A crossroads where all drivers must yield to one-another do not make sense and is a sure recipe for an angry gridlock. Updating the roadsigns corresponding to changes in rules of circulation is a clear example of cognitive nice maintenance.

Some interesting literature on cognitive niche maintenance can be drawn from a related but not overlapping field, namely the research on chance discovery (Bardone 2011). As defined by Oshawa and McBurney, a chance is a new event or situation conveying both an opportunity and a risk in the future, depending on the agent who detects it (Oshawa and McBurney 2003).[2] Similarly to affordances (Gibson 1977), chances belong to a relational ontology: they may or may not be discovered and used

[2]In spite of the terminology, the notion of chance does not focus on its aleatory dimension, but rather how the agent's knowledge and skills will enable her to perceive that opportunity and act upon it to achieve positive outcomes (Bardone 2011).

by an agent because of her capabilities, her cognitive endowment, her knowledge and her attention. Abe (2010) further illuminated the notion of chance curation, that is very akin to what we mean by cognitive niche maintenance. Abe remarks that the "main task of curator is a curatorial task, which is multifaceted. Curator comes from a Latin word 'cura' which means cure [and care]. Then originally it was used for a person who takes care of a cultural heritage" (p. 794). The word was subsequently applied to the person who takes care of data in IT settings:

> They use 'data curation' because they think data have value. Not only for keeping data but also usability of data for the public, they use the word 'curation'. Actually, most of data are neither art works nor archaeological artifacts. However, is important to view data from the aspect of what should be preserved. (p. 795)

The similarity between affordances and chances is sufficient for our discourse. Since Gibson defined a niche as "a set of affordances" (Gibson 1979, p. 128), we can extend this definition to see a niche as "a set of chances": it seems therefore sensible to employ the expression *cognitive niche curation* for activities of cognitive niche maintenance and updating. It easy to see that ignorance is a major concern for niche curation activities.

2.2 The Meaning of Ignorance in a Niche Constructionist Perspective

In her thorough philosophical and cognitive analysis of ignorance, Arfini (2019) describes two major ways in which ignorance affects cognitive niche construction:

> Ignorance affects the development of cognitive niches in the sense that if a particular lack of data or information is shared by the occupants of a cognitive niche, then they also share epistemic and ignorance bubbles about it: it is virtually impossible for them to pop their bubbles without the injection of data and information external to the original cognitive niche. [...A] way to diffuse ignorance in cognitive niches is through the limitation of the available affordances for the epistemic agents. The establishment of taboos (as ways of conducts that are forbidden in a particular community) does not only affect the lack of propositional knowledge that the inhabitants of certain groups share, but also the creation and perception of affordances that involve the objects and concepts banned. (pp. 145–146)

Ignorance seems to concern cognitive niche construction in two major ways. First of all, quoting Tooby's definition, cognitive niches are about "the ability to create and maintain cause-effect models of the world as guides for prejudging which courses of action will lead to which results." Ignorance, as the lack of knowledge or the presence of faulty/corrupted knowledge, entails the impossibility to maintain effective cause-effect models, which will result in a poor prediction of the effects of our behavior.

Let us go back to traffic circulation, and think of the difference between right-driving and left-driving countries (UK, some actual and former Commonwealth countries, and a number of Asian countries including Japan and Singapore). Comic movies abound with instances of British drivers landing in another country and driving on the wrong side of the road. Apart from the caricature, there is a reason why this is more likely than the other way around: in the United Kingdom, circulation is very verbose. Fully embodying the cognitive distribution at stake in cognitive niches, the UK road system is covered in writing (on the pavement, on signs before roundabouts etc.) constantly advising drivers which way they should go. In comparison, the US road system is much less verbose, there are fewer written indications constantly reminding you on which side you should be driving, and it is therefore easier for an absent-minded driver to err on the wrong side of the road. The cognitive externalizations are less verbose and require more knowledge, for instance internalizing that the yellow line separates directions while the white line separates lanes. Not being able to pick this ecological sign and regulate one's expectations and one's behavior accordingly show the risk brought about by ignorance in a cognitive niche. The group, or particular individuals, living a given cognitive niche warrant for the dissemination of effective cause-effect relationship to be activated as the relevant prop is perceived.

As amply discussed in Magnani's Understanding Violence (2011), cognitive niches are a socially distributed phenomenon. In other words, cognitive niche construction has a transversal dimension, corresponding to the relationship between cognizants and their cognitively enriched environment, and a horizontal dimension, binding together the cognizants partaking of a same niche as a *group*. What is the further problem with instances of behaviors induced by ignorance? The incapacity to select or activate the appropriate cause-effect model to guide one's behavior may signify two epistemic failures: either the appropriate cause-effect model is lacking, or the relevant prop is not recognized as such. Both can also be the case. Let us keep traffic as an example. In the UK virtually all roundabouts are…roundabouts! The rule embedded in the roundabout, also reminded by the road signals, is that incoming drivers must yield to traffic that has already engaged the roundabout. France, as usual, is peculiar: although over the past decades the country has embraced roundabouts as substitutes of rural crossroads (with the aforementioned yield rule, which is known in Southern Europe as "French roundabouts"), in a city one can find massive roundabouts, such as the Place Charles de Gaulle (formerly Place de l'Étoile) or Place de la Bastille in Paris, that are actually…crossroads, where the rule is "yield to all vehicles coming from the right". Unaccustomed drivers display two (erroneous) behaviors: either they break before entering the roundabout, or they fail to break for vehicles entering the roundabout. Such errors are clearly triggered by a form of ignorance: not so much the lack of the cause-effect model, though, but the incapacity to recognize a prop as relevant for its activation. In other words, unaccustomed drivers are aware of the different driving behaviors required at a crossroad and at a roundabout, but they are unable to recognize that certain roundabouts are actually crossroads. Needless to say, such a behavior will either cause scorn, or anger (depending on the gravity of the consequences)—on top of being a way to point out "foreigners," who do not fully partake of the same cognitive niche as locals.

Cognitive niche construction is about the persistence of the cognitive niche over time, which allows the availability over time of the benefits brought about by the cognitive reframing of the environment. As stated by Clark (2005), the basic and most overwhelming cognitive niche is the one structured by human language, the first sharable cognitive scaffolding human beings deployed over the word. Before the development of written language, or in predominantly oral cultures, the safeguarding of the copy fidelity of the information constituting the cognitive delegations was of paramount importance for the performance of the group. This lets us appreciate the importance of the integrity of the environment-niche-group triad, which eventually sparked extreme coalition enforcement practices (Bingham 2000; Rohwer 2007). Ignorance of some of the knowledge vital to the cognitive niche can become a hazard for the whole community, hence ignorance in general might have constituted as something against which we developed a powerful aversion. The immorality (and hence punishability) of the ignorance concerning basic elements of group membership such as language can be traced in two interesting instances: on the hand, the negatively-connoted word *barbarian* comes from the Greek *barbaros* and would refer to those who do not speak Greek, as people speaking other languages would be perceived as "babblers" (the word is originally onomatopoeic); on the other hand French , which underwent a massive reformation under Napoleonic rule in the early nineteenth century forsaking the tolerant difference between private and public orthography, displays a similar hostility against the failure to demonstrate accuracy in such a group-binding knowledge as language. The French word for a language mistake—be it orthography, grammar, or syntax—is *faute*, which is the same word employed for a moral infringement: not knowing your French makes you *faulty*.

The fact that cognitive niche constructors are connected among each other and not only with the target domain of the niche explains why ignorance has been the object of severe reprimand, no laughing matter, until a given niche was robust enough to become resilient against more or less extreme instances of ignorance: such robustness is a characteristic of the relationship between the niche and its target domain, but also among the members of the group living the niche.

3 Humor and Ignorance

3.1 Ignorance and the Incongruity Theory of Humor

The Incongruity Theory of humor can be seen as an epistemological understanding of humor. Now commonly accepted, it emerged in the eighteenth and nineteenth centuries in contrast to a previous moral-based theory (the Superiority Theory, which claimed that humor arose from a feeling of superiority between the subject and the object of laughter) and a physiological one, the Relief Theory (describing humor as a discharge of nervous energy).[3] The essence of the Incongruity Theory is that humor is caused by the perception of two or more incongruous elements. It was first introduced by James Beattie, who wrote that:

---

[3]Aristotle hinted at the Incongruity Theory of humor his *Rhetoric*: it was not so much about the cognitive reception of incongruity as humor, as an advice on how to generate hilarity by setting up some expectations and then violating them (Aristotle 2018).

> [Laughter is caused by] two or more inconsistent, unsuitable, or incongruous parts or circumstances, considered as united in one complex object or assemblage, as acquiring a sort of mutual relation from the peculiar manner in which the mind takes notice of them. (Beattie 1779, p. 320)

Although Kant took side with the Incongruity Theory to explain humor and laughter, he did not employ the term incongruity (Kant [1790] 2008), we owe to Schopenhauer the modern fortune of the Incongruity Theory, as he expressed it in the first volume of *The World as Will and Idea* (Schopenhauer 1818/1844 [1907]):

> The cause of laughter in every case is simply the sudden perception of the incongruity between a concept and the real objects which have been thought through it in some relation, and laughter itself is just the expression of this incongruity. [...] All laughter then is occasioned by a paradox, and therefore by unexpected subsumption, whether this is expressed in words or in actions. This, briefly stated, is the true explanation of the ludicrous. (I, 13)

The Incongruity Theory of humor provides a compelling explanation for what is humorous and causes laughter as a mismatch between two cognitions, namely between what is actually the case and what we would expect to be the case. In this sense, humor displays a relational ontology as well. Nothing is intrinsically funny, and you must have some kind of expectation in order to grasp something as humorous. Certain things are nearly universally ludicrous, such as posterior bodily noises, because in most cultures they are considered as gross and not to be publicly let out. Other things are categorized as inside jokes, such as astronaut humor (*"I don't get it" "Oh, you had to be there!"*), or parodies: the more you are acquainted with *Star Wars*, the more you can appreciate Mel Brook's *Spaceballs*.

One of the main objections to the Incongruity Theory of humor is the matter-of-fact statement that not everything incongruous is ludicrous. Sure, this can bring us to explore the degrees and variations of humor as it develops into irony, cynicism, and bitter humor, but it is safe to say that, generally speaking, humor can hardly make room to accommodate certain incongruous events: for instance, the scene of a child knocking on a door to receive her Halloween treats and getting shot at instead is incongruous but not humorous for most human beings. Michael Clark's addition to the Incongruity Theory concerns the element of enjoyment: he suggests that "amusement is *the enjoyment of (perceiving or thinking of or indulging in) what is seen as incongruous, partly at least because it is seen as incongruous.*" Very few people normally *enjoy* the scene of a child getting shot across a front yard.

What does the Incongruity Theory tell us about ignorance and why it can be perceived as funny? Some instances of ignorance can be rightfully seen as an incongruity between what is expected to be, i.e. knowledge, and its actual absence or lack of accuracy. Such discrepancy becomes apparent through an ignorant performance or an ignorant utterance, either to the self or to other people. Becoming aware of one's or someone else's ignorance may or may not be ludicrous. Imagine a few situations:

- I always fly out of La Guardia airport and never from JFK. One day I am booked on a flight departing from JFK, but I still head out to La Guardia out of habit and realize my local ignorance (lack of accurate knowledge) when I get to the check-in desk. If my travel is not important, and/or I have time to rebook a flight or journey to JFK, I can laugh at my goofiness and the whole incident will become a funny story to share. If I am traveling for an important matter, say a job talk, and I cannot amend my mistake by any means, or that ends up costing me several hundred dollars, then it is not so funny anymore (and I will draw commiseration mixed with reprimand).

- A first-year medical student failing to locate the spleen and the gall bladder on the appropriate respective sides of the human body can provoke laughter (no matter how scornful). A surgeon displaying the same ignorance in the ER is no laughing matter.
- If a college student misquotes Caesar as having said "I came, I saw, I went" is ludicrous. The same sentence, confidently uttered by a full professor of Roman history grossing more than a hundred thousand dollars a year and heading the department would rather stir outcry and indignation.

Everyone's experience abounds with phenomena of ignorance that came across as hilarious according to the Incongruity Theory of humor, and with other events that—albeit displaying a level of incongruity—were not deemed to be ludicrous. Can the theory of cognitive niche construction help us tell the difference between these matters?

3.2 *Viscosity: Humor, Benevolence, and Harm*

Lahti and Weinstein introduced a very interesting notion regarding the evolution of moral judgement: *moral viscosity* (Lahiti and Weinstein 2005). Theoretically speaking, the infringement of a moral principle may trigger the punishment of the transgressor, or the forsaking of the principle: one of those options should obtain, because the whole point of morality is its enforcement. Lahti and Weinstein's claim is that the additional possibility of forgiveness highlights the "viscous" quality of moral principles: under certain conditions (for instance, if the violation is not too frequent or systematic), their failure does not entail their negation. Is a similar viscosity a prerogative of cognitive niches as well, and how would it affect ignorance and humor? Cognitive niche construction, as we saw, is about cognitive delegations that extend out of individual minds. For a cognitive delegation to be an activity of niche construction, it must afford a certain degree of persistence. As observed above, spoken language is the most basic activity of niche construction, but the persistence of an oral culture is far different with respect to that of a written culture. The former rests on patterns of neural activations, on acoustic waves, and on the nervous wiring connecting the ears and the mouth with the brain; the latter's cognitive delegation is mediated by a fully external physical support that can be always accessed. Ignorance comes to acquire different meanings in either case: when forgetfulness or epistemic corruptions affect a niche whose persistence is akin to that of an oral culture, ignorance is always problematic inasmuch as it could affect, directly or by contagion, the relevant information that makes the niche profitable. In case of a niche whose persistence is not in question, an instance of ignorance (both in a cognizant or in a cognitive delegation) can be safely corrected and can also come across as funny, as in the case of the med student not having fully learnt his organs yet.

In a niche construction perspective, the degree of viscosity of the niche (which could also be defined as the degree to which the niche can tolerate ignorance) relates to whether ignorance can be accepted with humor or not.

An emergency room, the cockpit of a passenger plane, or a governmental data center are examples of poorly viscous niches. Instances of local or general ignorance in the handling of the niche have nefarious pragmatic consequences that may impact the survival of the group and of the niche. The incongruity between the desired and the actual knowledge is perceived but is not amusing.

Other cognitive niches, such as educational institutions or many of those in which everyday activities take place, are more viscous: the relationship between shared knowledge, externalized knowledge, and the direness of the consequences make room for a reception of the incongruity between actual knowledge and optimal knowledge as a source of amusement.

Drawing on the similarities between chance theory and cognitive niche curation, it makes sense to say that ignorance leading to the amused perception of incongruity signals the possibility of niche curation: in other words, ignorance perceived as an amusing incongruity points on the one hand to the mismatch between actual and optimal knowledge, and on the other hand the possibility to correct this mismatch—in a way that is positive both for the niche and the ignorant agent.

In a way, connecting ignorance, humor, and the perception of ignorance in the activity of niche construction seems to corroborate the Superiority Theory of humor, which goes all the way back to Plato and the Bible: spotting someone's ignorance makes us superior to that person, at least epistemically. But the feeling of superiority is not what is at stake here: first of all, spotting the ignorance as incongruity between epistemic states is an interesting phenomenon, insofar as it does not demand that we possess the knowledge that we deem lacking in another agent. For instance, it is possible that I recognize an incompetent (qua ignorant) doctor even if I (clearly) lack the medical knowledge that I think he is lacking. Secondarily, humor is indeed a form of aggression, but a self-defusing aggression: no matter how humiliating being laughed at can be, it is better than being insulted, physically hit, ostracized, or killed (Magnani 2011). Humor signals the incongruity and at the same time a degree of benevolence, or at least a lesser degree of intended harm against the ignorantly speaking or performing subject. Humor signals the possibility for a reparation of the niche construction activity, both at the level of the cognizant and of the available knowledge.

3.3 One Laugh Does Not Fit All

As it is empirically self-evident, reactions to instances of ignorance are not always the same and are not always trigger for humor. The same display of incongruous ignorance in the same niche may provoke different reactions in different witnesses. Clearly there is a gradient in viscosity on which the vibe of the reaction depends.

Cognitive niches, too, partake of relational ontologies: a niche is not just *out there*, rather it depends on how cognitive agents interact with it. Furthermore, the cognitive niche gives itself differently to different agents, according to the relationship one entertains with the niche: the most relevant factors seem to be one's *knowledge* about and *stance* towards the niche.

Let us consider the stance first: one can be a regular user of the niche, appointed and recognized gatekeepers, informal and unrecognized gatekeepers. Appointed and recognized gatekeepers act as enforcers of the niche, they curate the relationship between its knowledge, its target domain, and its users: they include instructors, teachers, but also enforcement officers, managers, family-members and anyone with a legitimate authority over someone else at a certain moment of their life. It's important to distinguish them from self-appointed and unrecognized gatekeepers, people who feel the obligation to protect the niche without being called nor appointed to: think of grammar nazi, vigilantes, zealots, activists, reformed-sinners and many, many others.

The other relevant aspect impacting the reaction in the case of an instance of ignorance is knowledge: while little knowledge can trigger alarm in view of the possible consequences of the epistemic failure, extensive knowledge of the niche and its dynamics will provide a clearer view of the possible outcomes of the instance of ignorance, its possible solutions, and will not only see the pragmatic halt imposed by the ignored matter, but also what ought to be in lieu of the ignorance, and hence better perceive the incongruity—and with incongruity, the humor.

Who, then, is better able to thrive in the viscosity of the niche? Either people who know how resilient a niche is, or people who know they can rely on knowledgeable people to correct an occurrence of ignorance. They can either perceive the incongruity, or trust the wishful belief that someone else can. People with little knowledge of the niche, especially if unable to trust the knowledge of fellow members as higher than theirs, might also perceive the niche as less robust and hence lack the elbow-room for a humorous reaction before ignorance, rather advocating a violent, retaliatory response against who, through ignorance, seems to threaten the survival of the niche.

Self-appointed enforcers appoint themselves because they believe the cognitive niche they care about is in danger and in need of rescue: in this view, the incongruity theory of humor cannot even be activated because the pragmatic danger is too dire and the reaction must be swift and stern. Even a strong niche is perceived as weak, and weak niches do not make room for forgiveness. Ignorant and unconcerned members must be rebuked, corrected, or eliminated (Magnani 2011).

Appointed enforcers, conversely, can perceive the humorous incongruity, but might nevertheless be prompted into a not amused reaction because of their very role. A teacher might find the mistakes of her students hilarious, and still fail their exams so they can learn: similarly, a police officer can empathize with the ludicrous account of a speeding driver, and yet cite him for his reckless behavior who might have endangered those he shared the road with. Sometimes, parents have a hard time not laughing at those misbehaviors of their children's that are caused by a lack of knowledge, but they still scold them to educate them about the correct behavior that would protect them in a less safe situation.

This analysis is parallel, to a certain extent, to the one that Nietzsche drew in his *Genealogy of Morals* (Nietzsche 2007). The German philosopher contented that the richer a society grew, the more it could afford not to—and hence would not—prosecute and punish its wrongdoers. Although this explanation of why ignorance is sometimes a laughing matter is to a certain extent Nietzschean, it transcends the classical vision inasmuch as it does not only consider the robustness of the niche, but also the *perceived* robustness.

4 Conclusions: Private Ignorance, Public Laughter

Ignorance is often amusing to the extent that (1) it signifies an incongruity between the knowledge that the situation would presume and the ignorance that the situation actually displays; and (2) the possibility to solve this incongruity to the benefit of the niche and the ignorant cognizant. As we saw, not all incongruous things are amusing: think of poverty versus wealth, sickness versus health. Still, ignorance elicits amusement because no matter how hard it is to change one's or someone's mind, it is still easier to change that than the external world (for instance by increasing one's health or one's wealth). Ignorance is never a desirable state, but when its consequences are mild, the fact that it is perceived as amusing signals an opening towards its resolution as long as socio-cognitive and environmental resources available in the niche afford the curation process.

What is the relevance of this line of research for the public discourse? If the argument of the last section holds, it can be posited that a humorous reaction to ignorance, perceived as incongruity, obtains when there is enough knowledge and confidence to appreciate the incongruity between the current undesirable epistemic state and the desired one, together with knowledge on how to move from the former to the latter. Knowledge of the ideal state is necessary to perceive the discrepancy and find it humorous. When ignorance perceives ignorance, we are helpless. If the practical shortcomings brought forth by ignorance are perceived, but the ideal state is out of reach, different things ensue: fear, anger, hate, suffering—to quote the exiled philosopher of Dagobah.

In the history of humanity, increases in knowledge have always resulted in increased welfare. Increased welfare translated into increased happiness and increased happiness fostered the resort to humor as a reaction to ignorance, also while keeping correction as the ultimate goal. Conversely, epidemics of ignorance rarely boosted welfare and happiness, generalizing states of fear, suspicion, and incapacity to rely on the "fellowship" of other women and men.

If we translate all this in the contemporary political scene all over the world, can we hypothesize a connection between the anger and spite characterizing public discourse also in liberal democracies, and the degradation of the underlying epistemic situation? If so, studies as the one included in this collection are all the more vital to understand—and hopefully repair—that exquisitely human relationship between knowledge and civility, beyond the rule of offense and the curfew of barren political correctness.

References

Abe A (2010) Curation in chance discovery. In: 2010 IEEE international conference on data mining workshops. IEEE, pp 793–799

Arfini S (2019) Ignorant cognition: a philosophical investigation of the cognitive features of not-knowing. Springer, Cham

Aristotle (2018) Rhetoric. Hackett, London

Bardone E (2011) Seeking chances: from biased rationality to distributed cognition. Springer, Berlin/Heidelberg

Beattie J (1779) Essay on laughter and ludicrous composition, 3rd edn. Creech & Dilly, London

Bertolotti T, Magnani L (2017) Theoretical considerations on cognitive niche construction. Synthese 194:4757–4779

Bingham PM (2000) Human evolution and human history: a complete theory. Evol Anthropol 9:248–257

Clark M (1970) Humour and incongruity. Philosophy 45(171):20–32

Clark A (2005) Word, niche and super-niche: how language makes minds matter more. Theoria 54:255–268

Gibson JJ (1977) The theory of affordances. In: Shaw RE, Bransford J (eds) Perceiving, acting and knowing. Lawrence Erlbaum Associates, Hillsdale, JN

Gibson JJ (1979) The ecological approach to visual perception. Houghton Mifflin, Boston, MA

Iriki A, Taoka M (2012) Triadic (ecological, neural, cognitive) niche construction: a scenario of human brain evolution extrapolating tool use and language from the control of reaching actions. Philos Trans R Soc B 367:10–23

Kant I ([1790], (2008)] The critique of judgement. Oxford University Press, Oxford

Lahiti DC, Weinstein BS (2005) The better angels of our nature: group stability and the evolution of moral tension. Evol Hum Behav 2:47–63

Magnani L (2007) Creating chances through niche construction. The role of affordances. In: Apolloni B (ed) Knowledge-based intelligent information and engineering systems: 11th international conference, KES 2007, Vietri sul Mare, Italy, September 12–14, 2007, Proceedings, Part II, Lecture notes in computer science. Springer, Berlin/Heidelberg

Magnani L (2009) Abductive cognition: the epistemological and eco-cognitive dimensions of hypothetical reasoning. Springer, Berlin/Heidelberg

Magnani L (2011) Understanding violence. Morality, religion, and violence intertwined: a philosophical stance. Springer, Berlin/Heidelberg

Nietzsche F (2007) On the genealogy of morality [1887] (Ansell-Pearson K (ed), Diethe C (trans)). Cambridge University Press, New York

Odling-Smee FJ, Laland KN, Feldman MW (2003) Niche construction. The neglected process in evolution. Princeton University Press, Princeton, NJ

Oshawa Y, McBurney P (eds) (2003) Chance discovery. Springer, Berlin

Pinker S (2003) Language as an adaptation to the cognitive niche. In: Christiansen MH, Kirby S (eds) Language evolution. Oxford University Press, Oxford, pp 16–37

Rohwer Y (2007) Hierarchy maintenance, coalition formation, and the origin of altruistic punishment. Philos Sci 74:802–812

Schopenhauer A (1818/1844 [1907]) The world as will and idea (Die Welt als Wille und Vorstellung). Routledge and Kegan Paul, London

Tooby J, DeVore I (1987) The reconstruction of hominid behavioral evolution through strategic modeling. In: Kinzey WG (ed) Primate models of hominid behavior. Suny Press, Albany, pp 183–237

Wallach E (2015) Niche construction theory as an explanatory framework for human phenomena. Sinthese Online First. https://doi.org/10.1007/s11229-015-0868-0

Sensible Objects: Intercorporeality and Enactive Knowing Through Things

Tomie Hahn and J. Scott Jordan

Abstract This chapter integrates ethnographic techniques, cognitive science, and enactive theory to examine the phenomenology dynamics that emerge during spontaneous interaction during a newly developed practice called *banding*. Specifically, participants are *connected* to each other via large rubber bands. An enactivist analysis of participants' journals reveals participants undergo intense intercorporeal experiences with properties that: are disorienting; are multi-scale; conjure intercorporeal surprise and discovery; undergo patterns of change, in both groups and individuals; give rise to intercorporeal trust; and they entail intercorporeal shifts in identity. The paper analyses how these properties might reflect the intercorporeal nature of everyday experiences.

Keywords Intercorporeal · Embodied anticipation · Enactive · Sensible objects

> The question, Me? You? Us? becomes at once appropriate but totally unanswerable. As I cannot feel any differentiation—I am me.you.us. I feel it. Your pulse rocks the band that nudges my ankle that glides across the floor that takes your belly with it. And down to the ground you go. And then I. And you. And I. And you. us. I. continually. It is at the point where: We have become the same project. (Lindsay Vogt in Hahn et al. 2016: 153)
>
> The inadequacy of words emphasizes the body expressing. Perhaps this is why I have become drawn to making music with an instrument that has such an uncompromised relationship with the body—the sound only exists in relation to the body's essential process of breathing and the contours of the sound directly reflect the internal contours of the body. (Simon Rose in Hahn et al. 2016: 154)

T. Hahn (✉)
Department of the Arts, Rensselaer Polytechnic Institute, Troy, USA
e-mail: hahnt@rpi.edu

J. S. Jordan
Department of Psychology, Illinois State University, Normal, USA

T. Bertolotti (ed.), *Cognition in 3E: Emergent, Embodied, Extended*, Studies in Applied Philosophy, Epistemology and Rational Ethics 56,
https://doi.org/10.1007/978-3-030-46339-7_3

The two passages opening this chapter[1] are excerpts from two musicians' journals reflecting their experiences during experimental movement and sound improvisation sessions called *banding*, where participants are literally connected by giant rubber bands as they improvise. Lindsay and Simon's texts grapple with the very concerns this chapter explores—how to articulate the essence of new, multi-sensory experiences and creative expressions *with others through* a shared object.

Consider the variety of objects one physically encounters throughout a day. Some objects are manipulated, others support or house the body, and others aid communication, orientation in space, health, food intake, transportation, etc. While everyday objects assist bodies to relate to and extend into the exterior world (including other living beings, objects, and space), there are cases where select objects are chosen exclusively for their exceptional qualities to support creative expressivity and to convey that which is interior to the body, embodied cultural knowledge (Hahn 2007). In particular, performers and artists employ uncommon objects in the real world to serve as expressive tools—musical instruments, paint brushes, fans, or costumes, to name a few. Artists and performers hone specific objects and form unique relationships with them. Hahn refers to exceptional objects that artists employ for creative expression as *sensible objects* (2011). Sensible objects are particular things that extend the body/self to serve as a vehicle for expanded sensory means, to also enable a kind of informative sensory loop, and to creatively express one's self in that reality (through text, drawings, music, movement for example).

The looping back of what is sensed and what is expressed—the externalization of what is internal—has an added dynamic when considering that objects can be signifiers, carry cultural meanings, and even have a social life of their own (Bates 2012). In different cultural contexts and in different periods of time, sensible objects may shift their meanings and, as a result, the embodied practices that incorporate particular objects are culturally prescribed. This focus on objects and the role they play in affording interaction is important to the issue of intercorporeality, because quite often, when attempting to get a grip, as it were, on the essence of intercorporeal experience, researchers utilize observations and self-reports regarding everyday experiences that take place in well-known types of contexts. Of course, this practice is consistent with Merleau-Ponty's assertions regarding the pre-reflective, ubiquitous nature of intercorporeal experience (Meyer, Streeck, and Jordan, this volume). The unique contribution of the present chapter however, is that it describes a methodology for using objects (i.e., very large rubber bands) that afford groups of individuals the opportunity to couple their immediate actions, in real time, in ways they have never experienced before. That is, the objects provide a unique form of intercorporeality. Our hope is that by examining the responses people have to this immersive, novel experience, we can shed light on the nature of the pre-reflexive, ubiquitous intercorporeality individuals exist within every day of their lives, and the role that objects play in the generation and sustainment of such intercorporeality.

---

[1] Chapter reproduced from Hahn et al. (2017) with Permission from Oxford University Press through PLSclear. © Oxford University Press 2017.

1 Investigating Intercorporeity and Objects in a Well-Known Context

In a recent analysis of the role that objects play in everyday human interaction, Hahn and Jordan (2014) conducted an interdisciplinary analysis of pedagogical practices in a *nihon buyo* (Japanese traditional dance) studio. Specifically, they combined ethnographic analyses with recent findings in cognitive science that reveal that environmental (i.e. social and cultural) events that are consistently paired with ongoing actions are later able to prime those very same actions (Hommel et al. 2001). Hahn and Jordan utilized these findings to create a possible account of (1) the multi-scale patterns that emerge over time in a dance studio (e.g., the immediate movements taking place during a particular session, the patterns of movements learned over multiple sessions, the teacher-student dynamics learned over multiple sessions, and the rich *nihon buyo* pedagogical tradition that comes to be embodied in both the participants and the studio context itself), and (2) the larger scale cultural dynamics these emerging patterns ultimately give rise to and sustain.

Within the above-mentioned analysis, the *nihon buyo* studio was conceptualized as a culturally relevant, practice-sanctioned object (i.e., an external scaffold) that afforded individuals a context within which to learn (i.e., embody) and sustain multiple scales of cultural practice. The object (i.e., the studio) in this case, afforded the development of multiple scales of *embodied anticipation* (Hahn and Jordan 2014; Kinsbourne and Jordan 2009). That is, it afforded the development of implicit, embodied expectations about the relationships between movements and the sensory effects they reliably produce. As an everyday example of embodied anticipation, when walking down a flight of stairs while talking with a friend, one is completely unconscious of the fact that one is continuously generating unconscious expectations about where one's feet are; that is, until one starts to fall. Then, the moment one's unconscious movement anticipations are disrupted, the disruption moves into conscious awareness, and one reorients one's current focus in the current context and, instead of talking, attempts to prevent the fall. It is this moment of conscious surprise that reveals one was unconsciously generating movement anticipation for the foot prior to the disruption (Jordan 2013). As regards the *nihon buyo* studio, Hahn and Jordan (2014) propose that the studio, as an object, affords individuals the development of embodied anticipation at many different, nested time scales (e.g., where a student should position her body in relation to a teacher at the start of a session, how the student should carry her body in relation to the teacher's during the session, how the teacher should move her body in relation to the student's as the student progresses in her learning of a particular dance). The emergent intercorporeal experience is also multiscale, such that at any given moment, movements within the context give rise to multi-scale experiences of being embedded within something beyond the immediate moment; what one might refer to as a *culture*.

2 Investigating Intercorporiety and Objects in a Novel Context

In the present paper we also discuss objects (i.e., scaffolds) and the means by which they are used to afford different types of interaction. Instead of discussing objects that afford the emergence and sustainment of everyday cultural practices however, we focus on the rubber bands used in *banding* because they afford individuals novel forms of interaction and, therefore, new forms of intercorporeity. *Banding* is an experimental practice for movement and sound improvisation that Hahn began developing in 2008 in which large industrial rubber bands literally connect participants with each other (see Fig. 1). Over the years *banding* has developed into a process for the exploration of playful ways of connecting, communicating, improvising, discussing, and writing about the senses, transmission, and relationships between individuals through things. The bands have become sensible objects, affording a variety of interactions.

Fig. 1 *Banding* session, Marlboro College, Vermont, 2011

2.1 The Methodology of Banding

Banding interactions tend to involve 10 or more participants, 10 or more very large, palette bands, and a large open space. In Round 1, what is called the *sensitizing level*, participants work in pairs, place a band around an arm, elbow, torso, or shoulder, and then loop the same band around their partner's arm, elbow, torso, or shoulder. Participants begin the *banding* interaction by *turning away* from each other and exploring various levels of subtle movements. They are instructed to start slowly and to pay attention to whether or not they can sense their partner's presence through micro-movements. They are also instructed to close their eyes and reflect on the qualities of movement they can sense, their changes in body position, and the speed of their movements. They are then encouraged to improvise and explore new interaction possibilities as they gain trust and awareness of their collective movement possibilities. After a period of improvising for 10–20 min., participants are asked to transition from *banding* to writing without conversation.

In Round 2, what is referred to as the *safe play, eyes open* level, participants work in groups of between three to five people and attach bands to the torso or any appendage (legs, arms, or neck). They are then instructed to face in any direction, keep their eyes open, and explore the dynamics of the group in flow. They are told to begin moving slowly and to gradually increase and vary their speed, pressure, direction, or the appendage to which the bands are attached. At a certain point the leader combines dyads so that larger groups form, increasing the complexity of relationships and interactions. Participants are asked to pay special attention to the mechanics of pull when there are several people connected in a group, and whether or not there are any differences between Round 1 and 2.

In Round 3, what is known as the *sound banding* level, participants are asked to add sound to their movement. Participants are asked to reflect on if and/or how the addition of sound changes the experience, and whether or not it changes the nature of the improvisation. They are then instructed to go off by themselves and engage in freewriting and/or freedrawing regarding the experience. Upon the completion of the freedrawing/freewriting interval, participants are brought together for a group discussion.

2.2 Measurement and Interpretation Issues in Banding

When conducting ethnographic research, especially when observing group dynamics, it is constructive to ask what participants' journal reflections might contribute to the research. Do such texts provide a view into the world of experience and meaning making? Do journal notes externalize that which is internal to the body? What role do textual reflections play in perception and the processing of sensory information to reveal aspects of consciousness, of intercorporeality? There is no doubt that improvisers physically experience *something* collaborative and intercorporeal through *banding*—the question is, what aspects of the processing of experience and sense-making does journaling reveal?

To be sure, the question of observing one's own perceptions of self and those of another has long occupied psychological, cognitive, and philosophical endeavors. In *Reflection and Interrogation* Merleau-Ponty questions:

> How are we to name, to describe, such as I see it from my place, that *lived by another* which yet for me is not nothing, since I believe in the other ... Yet should the voice alter, should the unwonted appear in the score of the dialogue, or, on the contrary, should a response respond too well to what I thought without having really said it—and suddenly there breaks forth the evidence that yonder also, minute by minute, life is being lived: somewhere behind those eyes, behind those gestures, or rather before them, or again about them, coming from I know not what double ground of space, another private world shows through, through the fabric of my own, and for a moment I live in it; I am no more than the respondent for the interpellation that is made to me. (Merleau-Ponty 1968: 10–11)

In *The Visible and the Invisible* Merleau-Ponty prioritizes *visual* sensibilities, running with, and then problematizing, the *seeing is believing* trope. Yet, after the recent *sensory turn*, how do we broaden the question of perceiving the realities of others, if only enroute to an exploration of intercorporeality? Suwa (2008) asserts that, "Verbalization of thoughts is a kind of externalization. It seems natural that meta-cognitive verbalization may encourage discoveries of unheeded variables of one's own body ..." (p. 141). Phelan (1993) on the other hand, problematizes the notion of a straightforward *analyze and reveal* relationship between performance (i.e., *doing*) and writing about what one did. For Phelan, writing is, itself, a type of *doing* (i.e., performance):

> Performance implicates the real through the presence of living bodies ... To attempt to write about the undocumentable event of performance is to invoke the rules of the written document and thereby alter the event itself ... The challenge raised by the ontological claims of performance for writing is to re-mark again the performative possibilities of writing itself. (Phelan 1993: 148)

Participants as embodied 'fieldsites.' In *Sensational Knowledge* Hahn (2007) theorized that the dancers that she observed and interviewed in a Tokyo dance studio were embodied fieldsites—sites of embodied experience. Hahn and Jordan (2014) referred to such embodied knowledge as *embodied context* to reflect the fact that, over time, the dynamics of moving bodies in a given context (e.g., a dance studio) come to entail the multi-scale dynamics the student (and the teacher) have had to embody over time in order to successfully learn and sustain a specific dance, learn and sustain a particular student-teacher interaction dynamic, and, ultimately, learn and sustain a cultural practice. It is because the cultural knowledge embodied within each dancer is not displayed unless he or she dances, speaks or writes about their embodied experiences, that Hahn considers dancers' bodies *as* fieldsites. The challenge is to investigate ways to reveal individuals' embodied knowledge and experiences, much like one would visit a fieldsite to explore the nature of a particular cultural environment. Depending on the project, methodologies such as interviewing, videotaping, sketching, and creating fieldnotes, are viable approaches to exploring such embodied fieldsites. Similarly, the journal entries of the *banding* participants (e.g.

Simon Rose's second quote opening this chapter) offer a glimpse of their personal *fieldsite*, identity, and their struggle to put words to an embodied experience.

The multi-scale nature of intercorporeality. The idea that writing is a different type of action that might actually change the experience it describes is consistent with the previously mentioned analysis of pedagogical practice in a *nihon buyo* dance studio. Again, Hahn and Jordan (2014) described how the persistent pairing of the student's own movement effects (i.e., the multi-sensory feedback derived from moving) with (1) the movements of the instructor, (2) the different phases of interaction during a single pedagogical episode, and (3) the overall context of the dance studio, gave rise to multi-scale intercorporeal experiences, what they referred to as multi-scale embodied anticipations. Though *banding* interactions also contain multi-scale dynamics, the different levels of scale are somewhat different. Common to both *nihon buyo* and *banding* is the level of immediate sensory effects generated by each individual participant, and the ability of these effects to be paired with the movement commands that produced them and, as a result, give rise to new embodied anticipations.

As regards larger-scale anticipations, the fact the *nihon buyo* students engage in multiple pedagogical episodes that take place over many years affords them the opportunity to embody expectations regarding larger scale patterns such as the sequence of events during a single session, or the change in instructional practices over the course of a year. While *banding* participants may meet several times during a multi-day workshop, they do not participate in the larger scale interaction patterns experienced by *nihon buyo* students.

Despite their lack of long-term *banding* interactions, *banding* participants nonetheless engage in multi-scale experiences. Specifically, during a *banding* episode, the *banding* itself provides a feedback loop that is physical and real. After *banding* sessions, participants enter into yet another loop in which they process their experiences by creating written and drawn artifacts that in turn encourage meaning- and sense-making. The activity of *banding*, creating artifacts, and sharing experiences function as a multi-scale feedback loop between the individual body, the bodies of others, the environment, and the artifacts (e.g., texts and images), group members externalized into their environment (i.e., shared with others), and provided the opportunity to embody the multi-scale nature on intercorporeality in a novel context. Analysis of the ethnographic data will let us know whether or not the writing actually changed the intercorporeal nature of the *banding* experiences.

Interpreting ethnographic data through the enactive perspective. Since 2008 Hahn has collected hundreds of written reflections and drawings from *banding* improvisers, in addition to interviews, email correspondence, and her own fieldnotes. During that time, she has found that the texts lend themselves to interpretation via the enactivist theoretical perspective. Enactivism offers an examination of intercorporeal experiences such that "social understanding … arises in the moment-to-moment interaction of two subjects" and includes "components such as bodily resonance, affect attunement, coordination of gestures, … and others" (see Fuchs and De Jaeger 2009: 465, 466).

Outlining the core difference between enactivist and traditional views of cognition Di Paolo et al., offer the following:

> Many influential theories in cognitive science make use of the idea that value of meaning is some information apprised by an internal module within an agent's cognitive architecture, whereas in an enactive perspective, meaning is inseparable from the whole of context-dependent, life-motivated, embodied activity, without being at all a hazy concept beyond the reach of scientific understanding. (Di Paolo et al. 2010: 35–36)

Central to the enactivist position is the notion that interaction naturally entails *sense making*.

> In everyday life, we experience the world in value-laden terms. This fact is hard to avoid and has been the subject of much philosophical debate. For enactivism, value is simply an aspect of all sense-making, as sense-making is, at its root, the evaluation of the consequences of interaction… (Ibid.: 45)

Evaluation of the ethnographic data will inform us of the types of sense-making that emerged during the course of the *banding* episodes.

2.3 Intercorporeal Patterns Revealed by **Banding**

In order to garner a better understanding of the ethnographic method she utilized to investigate *banding*, Hahn observed participants as they wrote in their journals and noticed participants painstakingly attempting to write, some frustrated and others enthusiastically writing many pages non-stop. Participants often remarked to Hahn that the journal writing helped them to make sense of the *banding* session. The following is an example of a journal text written after a *banding* session in Guelph, Canada, by Catherine Lee:

> It was amazing how it developed. I ended up with my eyes closed the whole time and it really felt as if I was inside the movement. Similar to how I feel when I am playing my instrument and am inside the sound or tone. I could not tell if I was moving or being moved. Such an incredible feeling of trust and beauty: the four of us moving together.
>
> The introduction of improvisation with an object was novel to me. The elastic allowed me to have much more freedom and to really play. It was much easier for me to get past inhibitions of being wrong or what people would think of me as we were all playing together. It has made me realize that when working with something that we have preconceived ideas about, such as an instrument, it can be difficult to enter into play; using an object that is new allows a freedom that in turn opens new ways of experiencing relationships and situations.
>
> During the explorations with movement and sound, the two felt inexplicably linked. The sensation was similar to how I had felt earlier during the *banding* when I was not sure if I was moving or being moved. It felt like a manifestation of the same thing, though rather than linking me with others this experience joined two systems that coexist within me. I was not sure if my movement was informing the sound or the sound the movement. They were one and the same, inseparable yet at the same time each with their own agency creating a dance of sorts. (Hahn et al. 2016: 156)

Lee's journal writing provides a characteristic case of how participants processed *banding* experiences through the activity of writing. Lee raises several insights from her *banding* experience in her text, including her impression of being *inside the movement* and comparing this new feeling to making music. She writes about the sensibility of being part of a group in both a physical and psychological way and links the group bonding to the bands themselves. She associates her awareness of intercorporeality and group expressivity with the bands as physical objects. Here Hahn's notion of *sensible objects* manifests as a feedback loop supporting the intercorporeal experience. The bands, as sensible objects, enable enactive group expressivity. Lee also writes about agency—of self identity with(in) the *banding* group.

During *banding* sessions Hahn noted the bands supported non-verbal communication over discussion, fostering instead tactile, visual, kinesthetic, oral/aural sensory modes of communication. Participants commonly described how time passed quickly or was suspended and that *returning* to the room was jarring. Journal writing and discussion of their experiences offered multiple means of processing the *banding* experience, but also a way to transition back to everyday modes of communication.

Consistent with Hahn and Jordan's (2014) interpretation of the emergence of the multi-scale intercorporeality that emerged in a *nihon buyo* dance across many pedadogical sessions in the same pedagogical context (i.e., the studio), the bands used in the *banding* interactions produced a physical, sensory feedback loop; the journals encouraged a larger-scale recursion on the *banding* experience; and the group discussions encompass all participants in a larger-scale, linguistically driven feedback loop. These multi-scale interactions influenced the overall intercorporeal experience of the individuals. As a matter of fact, the influence of public conversation was so powerful, the *banding* methodology was eventually changed. Specifically, after a *banding* session individuals processed their experiences differently—some found discussion productive, while others felt discussion interrupted their personal realizations of the process. Hahn observed that spending time processing personal experience through writing *prior* to hearing what others *felt* or sensed, was vital to reduce influences between participants. For this reason, the method was altered so that there was no verbal discussion or written communication between participants in the initial rounds. However, at the close of the *banding* workshop, time was reserved for extended discussions, which once again, opened-up the larger-scale, verbal mode of interaction. Group discussions reinforced camaraderie, as the sharing of personal physical and emotional experiences became aligned in a verbal re-living of shared experiences. Hahn observed that the physical attachments that began with visibly apparent rubber connections also included a lively exchange of ideas that arose through discussion and a sharing of notebook writings and drawings. In short, the larger scale, verbal interactions that followed *banding* episodes, influenced the smaller scale intercorporeal experiences participants had during subsequent *banding*.

Disorienting experiences. Hahn's observations of *banding* sessions, and interviews with participants revealed that *banding* offers an immersion in a unfamiliar sensory transmission context that enables a wide variety of physical and social coordinations. The unfamiliar banding practice provides a sense of physical disorientation that can often trigger social and psychological disorientations. Hahn found that during the first phase of a *banding* session disorientation soon transforms to a *re*orientation, through a swift growth in participants' embodied anticipatory knowledge, supporting the notion of *orientation through disorientation* (Hahn 2006, 2007) that can occur when people are in unfamiliar contexts. *Orientation through disorientation* refers to the initial experience of disorientation that an individual encounters, followed by an orienting, enculturating stage as the participant compares the new information with past experiences, and finally how the participant incorporates the previously unfamiliar sensation into their realm of sensorial knowledge. In this way, the unfamiliar gradually becomes familiar; the individual (re)orients through disorientation and experiences growth.

During an initial disorienting encounter, such as *banding*, individuals are challenged because the bands severely alter the sensory effect of movement the participants have come to anticipate. Thus, even though they continue to generate the unconscious movement expectations associated with non-*banding* movement, these expectations are consistently violated, participants are disoriented, and need to generate new, contextually-relevant embodied anticipations. For example, participants noted being cautious at the outset of their first *banding* experience, yet soon felt they understood how to coordinate safely with others, by counterbalancing or adjusting their movements fluidly.

On a physical level, disorientation during *banding* occurs when participants are in dyads and asked to sense the presence of their partner through a single band. The bands provide a buoyant quality so subtle movements of a partner can be sensed, such as breathing or a slight adjustment in stance. Participant descriptions of the phenomenon revealed that tuning into one's partner on such a subtle sensory level promotes empathy. Participants have conveyed that, over time, they feel a deep sense of connection to their partner and, even with their eyes closed, find they begin to anticipate even minute movements that enable them to adjust their own stance and easily maintain balance. Participants realize immediately that awareness of their partner is key, as their own physical stability and the group's stability depends upon a clear transmission of information such as movement, energy, timing, or weight.

Whether the empathy that participants convey in their journals and discussions is actually experienced by others, the consistent inclusion of empathy as part of an intercorporeal world within the bands is significant. The bands, as sensible objects, become conduits of empathy. As tangible and visual manifestations of the relationships between individuals, the bands afford empathy, as well as the responsibility and connectedness that comes with such coupling.

Banding as intercorporeal surprise and discovery. There is an unexpected, vibrant, presence that emerges during *banding*. It is unexpected, Hahn believes, because the bands are perceived as common objects connecting people. Participants do not expect common rubber bands to transmit so distinctly or deeply, the ephemeral nature of presence, of energy, of the sensory experiences during *banding*. Participants' journal texts and discussions effusively reflect the realization of intercorporeality—the surprise of connections. *Banding* participants readily articulated their processing of such experiences, such as Randall Ota in Colorado Springs:

> Afterwards I could very much feel everyone "in" me and "on" me, I could still feel their energy on me, within me, in my muscles, in my heart, in my feelings, in my sensibilities. This was a very pleasant sensation, and an experience that I think I had never had in quite that way, or probably in any way, for that matter. The closest thing I have experienced to *banding* would probably be dance improvisation, but *banding* put others "into" me in a much more visceral and definite way than dance improvisation. It was fascinating how, even if someone was several people away from me, I could still sense their movement, and even their "energy" for that matter. It was like we were all part of a big spider's web, and no strand could be plucked without everyone else sensing it, knowing it. *Banding* seems a concrete metaphor for a performance group working together as a team. No one can "out-pull" anyone else, everyone must match everyone else's energy, and this should be a good thing, since anyone can bring more out of everyone by increasing their own individual pull. (Personal correspondence 2012)

Ota imparted how *banding* enabled his heightened awareness of intercorporeality via tactile and enactive sensory connections with other participants' movements and "energy." When he noted sensing "several people away from me," Ota was referring to his experience of *banding* with someone who was banded to another participant. Ota, not directly attached to the participant on the oppose end, reported sensing their presence transmitted through bands and people. He succinctly points out interconnected themes that Hahn heard often—how *banding* heightens awareness of others and the bands illuminates relationships within the group, as well as how each participant's movements, however subtle, affect the whole.

Patterns of change in intercorporeal experience. From Sherrie Tucker's post-*banding* notes:

> At first I experienced choices.
>
> The choice to yield, to pull, the moments when I felt others yield or pull—then, we all became a body/organism and I stopped thinking in terms of what my own impulses were, which had been quite conscious—give, pull back, step over this one, lie down and roll with it—and tuned into what the organism was doing. Then, I noticed when subtle changes took place in patterns. I still anticipated snaps, the sonic-tactile ramifications of playing with rubber bands, and prepared to yield in these moments—but sensitivity to patterns felt different than choices, as listening is to sounding. It wasn't a passive experience to fold into the organism, but a different way of being. Senses attuned to bubbling and flowing, not to mention the smell of rubber, the unexpected pleasure of not consciously thinking through my steps and rolls in strands of rubber. (Hahn et al. 2016: 158)

Tucker's account of her "sensitivity to patterns felt different than choices" reinforces Hahn's observations of the transitional stages of *banding* where participants experienced a shift in orientation from the experience of being an individual within a group, to the experience of orienting *as* the group. Because the bands provide a visceral connection between participants, seeing and feeling the connections to others reveals apparent intercorporeal, relational aspects of the individuals within the bands, particularly the repercussions of one's actions. A feeling of responsibility to other participants quickly develops. Hahn notes that at this transitional stage participants begin to orient themselves to each other and develop a vocabulary of the *banding* practice. Because of this developing vocabulary of knowledge, participants can begin to anticipate the movements of others, while also providing stability or instability. George Blake wrote about this transition in a slightly different way:

> With "safe play, eyes open!" I lost familiarity with my body, regulated only by self-restriction. Instead, I was a co-creator of a web of sensibility. What others did became a constitutive part of my own experience. However, *banding* was a process of considering possibilities that involved testing boundaries. We had to create a solid base of rapport from which exploration and difference could exist. It was a performance inextricably linked to others. I couldn't formulate my next move without taking into account if and how new energy would impact the group. (Ibid.: 155)

Blake reveals how the bands expose participants' responsibilities to others to sustain the group—to be the group, acknowledging the intercorporeal nature of connectedness and emerging group identity.

The emergence of intercorporeal trust. At this point in the chapter we have arrived at a point of looping back. The theories of enactive and intercorporeal coordination meet, or merge with the *banding* case study examples. For example, Hahn observed the trope of band-as-support and feedback developed quickly, as participants realized that trust is vital within the intercorporeal looping of individuals within bands. The bands are strong enough that they can literally support a group's ability to stand, move, and balance. In some *banding* relationships the sense of support is subtle, while for others it can be dramatic. Francois Mouillot offers insights of a personal nature:

> As a 29 year-old suffering from cerebral palsy, my relationship to my own body had always been at the same time complex and very simple. On the one hand, my physical day-to-day activities have always consisted of dealing with physical obstacles in a never-ending search for balance (in a constant process of improvisation); on the other hand, I never believed or really felt that my body was itself an obstacle to leading an active life. In short, I never dreamed of having a different body.
>
> Given the particular limitations and possibilities offered by my body, I tend to have a fairly fixed idea as to what it can or cannot do. Control over my own physical balance is always a fleeting goal, and a temporary state of being.
>
> By contrast, the *banding* experiment struck me as an exercise in "letting go," and perhaps in a novel way for me, "letting go" of the urge (that I am so typically used to) to control my body. Through the gentle pushes and pulls inflicted on the rubber band, *banding* became an experience of listening. First, listening to my partners' suggestions as to what my body could do and where it could go, but also listening in the most straight-forward sense: Eyes-closed through most of the session, I felt very much aware of and guided by the sounds of our breathing, and of my body "cracking" and stretching. (Ibid.: 158–159)

The day Mouillot joined the workshop, the group banded in small ensembles. His way of finding, or negotiating, relationships with others influenced the larger group in profound ways, even for those who were not *banding* with him. Hahn watched Mouillot's trio playfully rolling on the floor in the sun, then his two partners stood to support him using facile *banding* techniques they developed. The "letting go" and trust Mouillot described in his journal demanded astute awareness, listening intently, and a closing of his eyes, revealing the depth of heightened sensitivities essential for the intercorporeal connection. The bands enabled support, movement, sound, as well as the connection with others.

Shifting identities and intercorporeal experience.

> My memories of the *banding* reveal a sensual experience—the feel of the tough, practical fabric of the band, the sound of my own breath and that of the three other people in my group, glimpses of the shifting shapes of the others, and time slowing down as this 4-headed, amoeba-like life-form half-oozes, half-flies through space. (Ibid.: 152)

Louise Campbell's quotation above offers insights into notions of a forming identity and world-making through enaction. Anthropomorphizing the group *banding entity* is common in participants' fieldnotes, such as Ota's "like we were all part of a big spider's web"—as participants attempt to make associations with newfound sensory encounters creates associations that form unique *banding* entities. In the second case, Campbell not only related her experience of becoming part of an imagined amoeba entity through tactile sensual encounters, but also through her sonic experience. She wrote about her changing awareness of time and space within this new identity. Campbell's reflections revealed the multi-sensory nature involved in processing her *banding* experiences, a making sense of the new experience within a group, or as a group. Intercorporeality, and a sense of an emerging identity, was enabled by the bands. Looping back to Vogt's opening quote in this chapter, the intercorporeal and interdependent nature of the *banding* experience reveals how individual identities are blurred and incorporated into a newfound group identity:

> The question, Me? You? Us? becomes at once appropriate but totally unanswerable. As I cannot feel any differentiation—I am me.you.us. I feel it. Your pulse rocks the band that nudges my ankle that glides across the floor that takes your belly with it. And down to the ground you go. And then I. And you. And I. And you. us. I. continually. It is at the point where: We have become the same project. (Ibid.: 153)

Here Vogt uses experiential description to explore the blurring and merging of separate, singular identities. This merging of identities and her questioning of new (disorienting) experiences maps into an enactivist sense- and meaning-making, a world-making emerging from *enacting a world. Banding* and such disorienting experiences shared with others in a group can potentially shift one's point of view of identity, but as the following post-*banding* correspondence reveals, it can also extend to one's enactive sense-making in everyday life experiences.

Relationships between time and the sensory experience of *banding* consistently arose in workshop discussions. Because *banding* is unfamiliar, it demands a lot of focus and energy to concentrate on the moment-to-moment personal experience while simultaneously maintaining an awareness of being part of a group. After *banding*, participants often voice that they continue to feel connected to each other and a feeling of swaying persists in their body. One participant compared the sensation to stepping off of a treadmill after jogging and continuing to feel the forward movement, despite being physically still. The sense of immersion within a group in *banding* can be overwhelming, as participants are building new physical memories alongside other participants. Jovana Milović wrote about how her *banding* experiences in the week-long workshop later affected her experiences in everyday life:

> The bands were both the precondition and the limitation of freedom and movement. Within the parameters of their elasticity and resistance, we tested the power and influence of the

> other bodies around us and became utterly and unmistakably *receptive*. This receptivity to the other—physical, literal, embodied—resonated through the remainder of our days together. We could not untwist from the bands that strung our thoughts into a dense constellation bright with creativity.
>
> Months later, I am on a subway car as it lurches forward, moving all the occupants in sync. Muscles strain to keep the bodies upright, still, appropriately dispersed. The corners of mouths curl upwards uncomfortably as, here and there, people wince. I'm sorry. I touched you. Curt nods forgive the transgression. The activation of the receptive realm is truncated by this place, as it is by so many of the places we inhabit.
>
> Watching people move in public places, I am fascinated by how effortlessly receptivity is foreclosed. After disturbances strangers silently renegotiate their space and retreat to their private sphere. Now, even shoulder-to-shoulder, we are disparate–each in a different sensory environment. (Ibid.: 160)

Milović's link between *banding* in the workshop and her real-world experiences points to an interesting characteristic of disorienting experiences. At the moment of a disorienting experience the encounter seems dramatic, but it establishes a reference point for similar experiences in the future—such as the subway ride. Consistent with Hahn and Jordan's (2014) multi-scale interpretation on *nihon buyo* pedagogy, the moment-to-moment sensibilities transmitted during *banding* often carries over—through associations, memories, physical impressions, and artifacts such as images and texts—to long-term embodied knowledge. That is, the long-term nature of Milović's experiential connection between her previous *banding* interactions and riding a crowded train, speaks to the multi-scale, dynamic nature of the intercorporeality. In short, all intercorporeality is enmeshed in a multi-scale network of meaning that emerges continuously and contextually out of one's sensorial interactions with persons, places, and objects.

3 Closing the Loop

The purpose of the present chapter was to conduct an interdisciplinary examination of the role that objects play in human intercorporeal experience. Instead of focusing on how certain objects are used to develop, afford, and sustain multi-scale embodied anticipation (Hahn and Jordan 2014; Kinsbourne and Jordan 2009) and, as a result, multiple scales of social practice, the present paper focused on a particular type of object (i.e., very large rubber bands) that afforded members of a group the opportunity to generate intercorporeal experiences that they have never experienced. Our assumption was that doing so would shed light on the ubiquitous, everyday, enactive nature of intercorporeality.

Textual analysis revealed that *banding* gives rise to disorientation as one comes to experience the movements of others in very detailed, kinesthetic ways one has never before experienced. Over the course of the *banding* interaction, one's movements, as well as those of the others, continuously give rise to intercorporeal surprise and discovery as the embodied anticipations one has developed over one's lifetime (up until the *banding* episode) find themselves continuously disrupted. Given that every single movement of every single member of the *banding* group is effectively *broadcast* through the *banding* network, one's access to the others plays out on a much finer, never before experienced, kinesthetic, enactive time-scale. In a sense, the notion of independent movement is gone. Every moment of movement entails the resistance brought about by being coupled with the other members in this unique kinesthetic field. As the interaction continues, participants experience shifts in identity from being a local, independent 'self' to being a member of a global, social field. Such enactive shifts in identity were often accompanied by the emergence of empathy and, what we refer to as *intercorporeal trust*. In addition, some participants reported the experience of 'aftereffects' when discontinuing the *banding* interaction, and some reported having a sense of 'group identity' for many days and weeks following the interaction.

While on the one hand, the data presented here might seem unique to *banding*, on the other hand, we propose they reveal important aspects of the emergence and sustainment of intercorporeality in everyday, enactive life. Specifically, we propose that the uniqueness of *banding* emerges out of the medium of the rubber bands and the fact that participants have no embodied anticipations regarding the sensory effects of being coupled with others in such a medium. These two facts provide participants the opportunity to experience the transition from conscious, reflective bodily experience, brought on by the disrupting influence of the bands and the need to consciously plan which movements are to come next, to what Merleau-Ponty referred to as pre-reflexive intercorporeality. This transition is possible because the repeated sensory effects of attempting to *move* in this unique medium, with a particular group of others, come to eventually be anticipated; not in a conscious way (e.g., "I am now anticipating that as I move my leg, I will sense the resistance of the other group members"), but in an unconscious way, what we have referred to throughout this paper as embodied anticipation (Hahn and Jordan 2014; Kinsbourne and Jordan 2009).

Such enactive, embodied anticipation is possible because environmental (i.e. social and cultural) events that are consistently paired with ongoing actions (e.g., the *feel* of a particular piece of clothing while performing a dance, the *feel* of the instructions a teacher gives via her hands and words during the performance of the dance, and the *feel* of the context within which the dance class is taking place) are later able to prime those very same movements, and come to be unconsciously anticipated during the generation of similar movements (see Meyer et al. 2017; Hommel et al. 2001). Given this relationship between planning, perceiving, and movement effects, *banding* participants come to experience their own movements *in terms of* the effects they repeatedly generate while moving through the constellation of effects entailed in the medium of the rubber bands. Given the effects of *others* are persistently, irreducibly entailed in the tension, direction, and force of the rubber bands, *others* become implicitly entailed in one's own movements and experiences, and the mode of this entailment is enactive.

In addition to being *other-relative* (because of the persistent presence of the others during *banding*), the resulting embodied anticipations are not conscious. Rather, as was stated earlier, and evidenced by participant Sherrie Tucker (see quotation in section entitled, *Patterns of change in intercorporeal experience*), the experience of planning what to do next transitioned from a conscious, *deliberative* process, to a seemingly unconscious, *receptive* process. That is, following a period of conscious practice, as movement dynamics were repeatedly paired with certain sensory effects (e.g., as I move in a particular direction, I sense the other's resistance), the sensory effects of moving came to be unconsciously anticipated during movement.

Given that unconscious, embodied anticipations are continuously generated for the entire body (Jordan 2013) we propose it may be the case that the constellation of embodied anticipations one is generating at any given moment constitutes an essential aspect of the phenomenally ubiquitous, pre-reflexive background of experience that Merleau-Ponty refers to via the notion of intercorporeality. By providing participants the opportunity to undergo the transition from experiencing themselves as a disoriented, enactive individual in the midst of others, to experiencing themselves as a fully-integrated, intercorporeal being, we propose we were basically allowing them to recapitulate the disorienting, re-organizing experiences that happen to all of us whenever our unconscious, embodied anticipations are violated by context (e.g., tripping while walking down a flight of stairs, behaving as if with friends and discovering a parent is present, or living in a different culture).

Given this notion, one might say that we tend to lose our awareness of our intercorporeal nature in daily life because the embodied anticipations we develop over time simply work from day to day and fade into our unconscious, pre-reflexive background. That is, as we move through the same contexts (e.g., the layout of our living space, the route we take to work, the layout of our work space) with basically the same people (e.g., family members, work colleagues, and friends) doing basically the same tasks (e.g., preparing for work, being at work, being at home at night) day after day, we embody these movement-event dynamics and they develop into unconscious expectations about being in, and moving through these contexts. As a result, we lose conscious track of our moment-to-moment, intercorporeal connectedness to *things* and *others* because the connections themselves are constituted of unconscious, embodied anticipations. In short, through refinement of embodied habituations over time, we become unaware of the necessary, generative role we ourselves play in our connection to *objects* and *others* every waking moment of every day. Such sentiment is reflected in the following, enactivist quotation:

> Though we cannot remember our experiences as infants, we were all nevertheless infants. We built our knowledge of movement and of the world on the basis of having learned our bodies and learned to move ourselves. We accomplished such learning by thinking in movement. (Sheets-Johnstone 2010: 172)

From Sheets-Johnstone's perspective, although we may not remember the disorienting nature of the experiences we had when first learning how to play patty-cake, how to tie our shoes, or how to read and write, the analyses revealed by the unique experience of *banding* indicate these experiences entailed surprise, shifts in identity, and the development of trust. Given these insights, the massive amount of human interaction that comes to be managed by the strategic employment of objects should come as no surprise. Street lights, textbooks, and the internet all constitute 'objects' we have created in order to sustain intercorporeal human culture. When looked at from this perspective, everyday life seems much more intercorporeal, and much more full of embodied trust than we tend to recognize.

Looping back, we close the way we began, with the impressions from *banding* participants:

> I am surprised by the connections,
>
> as our private thoughts are transformed,
>
> into a written elastic. (Catherine Lee in Hahn et al. 2016: 157)
>
> I can't help but think there is something absolutely aqueous about this whole thing. A particle system turns to a wave.
>
> We are so mixed up in each other that you, I, *we*, incur the complicated consequences of any action. Made unto, made within, this collective system. (Lindsay Vogt in Hahn et al. 2016: 153)

References

Bates E (2012) The social life of musical instruments. Ethnomusicology 56(3):363–395

Di Paolo EA, Rohde M, De Jaeger H (2010) Horizons for the enactive mind: values, social interaction, and play. In: Enaction: towards a new paradigm for cognitive science. MIT Press, Cambridge

Fuchs T, De Jaegher H (2009) Enactive intersubjectivity: participatory sense-making and mutual incorporation. Phenomenol Cogn Sci 8(4):465–486

Hahn T (2006) It's the RUSH': sites of the sensually extreme. Drama Rev 50(2):87–96

Hahn T (2007) Sensational knowledge: embodying culture through Japanese dance. Wesleyan University Press, Middletown

Hahn T (2011) Dancing with sensible objects (unpublished paper). Presented at the Society for Ethnomusicology conference, Philadelphia, PA

Hahn T, Jordan JS (2014) Anticipation and embodied knowledge: observations of enculturating bodies. J Cognit Educ Psychol 13(2):272–284

Hahn T et al (2016) Banding encounters—embodied practices in improvisation. In: Siddall G, Waterman E (eds) Negotiated moments: improvisation, sound, and subjectivity. Duke University Press, Durham

Hommel B, Müsseler J, Aschersleben G, Prinz W (2001) The theory of event coding (TEC): a framework for perception and action planning. Behav Brain Sci 24:849–937

Jordan JS (2013) The wild ways of conscious will: what we do, how we do it, and why it has meaning. Front Psychol 4

Kinsbourne M, Jordan JS (2009) Embodied anticipation: a neurodevelopment interpretation. Discourse Process 46:103–126

Merleau-Ponty M (1968) Reflection and interrogation. In: Wild J (ed) The visible and the invisible. Northwestern University Press, Evanston

Meyer et al (2017) Intercorporeality: emerging socialities in interaction. Oxford University Press, Oxford

Phelan P (1993) Unmarked: the politics of performance. Routledge, London

Sheets-Johnstone M (2010) Thinking in movement: further analysis and validations. In: Enaction: towards a new paradigm for cognitive science. MIT Press, Cambridge

Suwa M (2008) A cognitive model of acquiring embodied expertise through meta-cognitive verbalization. Trans Jpn Soc Artif Intell 23(3):141–150

Wild Relationality: The Skin Is Not an Epistemic Border

J. Scott Jordan, Alex Dayer, Jasmine Mason, and Vince Cialdella

Abstract The present paper questions whether or not metatheoretical approaches to cognitive science that utilize dynamical systems theory (DST) require a definitive means of demarcating the cognitive from the non-cognitive. As more researchers utilize DST as means to describe *relational* properties that emerge in an organism's interaction with its environment and then utilize relational properties as an account of cognition, we are forced to clearly specify (1) how relational properties differ from non-relational (i.e., intrinsic) properties, and (2) what, if any, roles the two types of properties play in constituting cognition. According to Wild Systems Theory (WST), this intrinsic-relational tension lies at the heart of current debates regarding extended cognition. WST asserts that if DST allows us to see cognitive phenomena in terms of *relational* properties, and not *intrinsic* properties, it might be the case that discovering the 'bounds of cognition' was never a problem in need of solving.

Dynamic systems theory (DST) has been used to explore cognition at the level of macroscale, animal-environment systems (Chemero 2011; Maturana and Varela 1980; Silberstein and Chemero 2015; Smith and Thelen 2003). Some argue this approach entails fundamentally flawed assumptions about what counts as cognitive (Adams and Aizawa 2001). If explanations of cognition can include features of the environment, we seem to have no way of delineating where cognition ends and everything else begins. The concern is that cognitive science needs a *definitive* 'mark of the cognitive' or a 'boundary of cognition.' Otherwise, cognitive scientists will have no clear-cut idea of what they are investigating.

The present paper questions whether or not metatheoretical approaches to cognitive science that employ the tools of DST require a definitive means of demarcating the cognitive from the non-cognitive. As more and more researchers utilize DST as a way to address *relational* properties that emerge in an organism's interaction with its environment and then utilize relational properties as an account of cognition, we are forced to clearly specify (1) how relational properties differ from non-relational (i.e.,

J. S. Jordan (✉) · A. Dayer · J. Mason · V. Cialdella
Department of Psychology, Illinois State University, Normal, USA
e-mail: jsjorda@ilstu.edu

T. Bertolotti (ed.), *Cognition in 3E: Emergent, Embodied, Extended*,
Studies in Applied Philosophy, Epistemology and Rational Ethics 56,
https://doi.org/10.1007/978-3-030-46339-7_4

intrinsic) properties, and (2) what, if any, different roles the two types of properties play in constituting cognition.

1 Embodied Cognition and the Relational Nature of Context

Although what counts as a body or not-a-body remains relatively unclear (Jordan and Ghin 2006), embodied cognitive science is often presented as a challenge to computational accounts that describe cognition as mechanistic manipulations of representational entities. However, some scholars don't think embodiment is nearly as new or radical as we might think it to be. In a recent paper, Raja et al. (2017) claim that embodiment, far from being a novel way of approaching cognitive science, has its roots in the transition away from Aristotlian physics to Newtonian physics. Embodiment can be seen as a reaction to a misguided attempt to describe the behavior of entities exclusively in terms of their intrinsic properties while ignoring the *relations* between entities that ultimately govern their behavior:

> For example, in Aristotelian physics, the behavior of a falling stone was explained in terms of its *substantial form*. Namely in terms of an intrinsic, inner feature that made the stone fall. In Newtonian physics, however, such a falling was explained in terms of a law (the Law of Universal Gravitation) that related the stone with another body (say, Earth) under some conditions. (2017, p. 12, emphasis in original)

The idea is simple: there is nothing intrinsic about the rock that causes it to fall. We can only explain why rocks fall in terms of relations between a particular rock and the environment in which it is embedded. Similarly, it is entirely possible that we may only explain some aspects of cognition in terms of relations between cognitive systems and the environments in which they are embedded. If we restrict our explanations of cognition to intrinsic properties, computations, or mechanisms, we run a similar risk to that of trying to explain why rocks fall in terms of their intrinsic, structural properties. In this way, we might see embodiment as an effort to be continuous with the rest of science by shifting the focus within cognitive science away from identifying intrinsic properties of entities to describing relations between entities.

One advantage of conceptualizing cognition in terms of relational properties is that it might actually move us away from our long-standing commitment to partitioning reality into objective and subjective properties. Famously, Chalmers (1996) argues that a complete functional/physical account of the world would not include an explanation of subjective experience (i.e., subjective properties). This is the "hard problem" of consciousness: the problem of explaining how and why subjective experience exists in a physical universe at all. For Chalmers, the hard problem exists because mind and matter are thought to be different ontological kinds; that is, the universe consists of distinctly subjective and objective properties. Interestingly, those who critique the subjective/objective distinction, often do so via appeals to relational properties.

1.1 Relational Properties and the Subjective

Silberstein and Chemero (2011) challenge the "hard problem" by conceptualizing subjective properties as a relational property. They are concerned that Chalmers' description of the hard problem forces us to either reject the existence of phenomenal experiences (eliminativism) or accept that phenomenal experiences are inherently otherworldly (dualism). Unfortunately, neither of these alternatives yields a satisfactory account of subjective experience. In a later paper, Silberstein and Chemero (2015) argue that we can deflate the hard problem of consciousness by "rejecting the idea that matter and mind are different to begin with" (p. 182). They then present a case for neutral monism: the view that the fundamental entities of the universe are neither physical nor mental, but something neutral between the two.

Silberstein and Chemero's version of neutral monism is distinct from existing accounts because it is "extended": that is, it draws on ideas from ecological psychology, embodied cognition, and dynamical systems theory to make the case that phenomenology is an indispensible feature of brain-body-environment systems. For the extended neutral monist, phenomenology is a feature of perceiving and actualizing affordances (i.e., relations between the perception-action capabilities of the organism and features of the environment). Because affordances are neither subjective nor objective, but rather "objective-subjective hybrids," they are exceptionally compatible with a neutral monistic framework. Thus, for Silberstein and Chemero, phenomenology is "neither in the head nor in the external world—it is fundamentally relational" (p. 190).

1.2 Relational Properties and the Objective

Rosen (2000), too, was dissatisfied with the distinction between subjective and objective properties. And he, like Silberstein and Chemero, proposed relational properties as a potential solution. However, instead of casting the *subjective* in terms of relational properties, he did so for the *objective*. To Rosen, those who were committed to the objective-subjective distinction seemed to be arguing that the objective world is a physical world devoid of consciousness, or a world independent of perceivers. In accordance with this traditional conception of objectivity, it is believed that the purpose of science is to uncover truths about this objective world (i.e., context-independent truths). In order to maintain a clear boundary between the objective and the subjective, scientists are led to reduce phenomena to simple mechanisms (or explain phenomena in terms of efficient causality). According to Rosen, recourse to mechanistic thinking safeguards science from observer-dependence, since "if a machine can 'do' something, that is prima facie evidence of its objectivity" (p. 83).

According to Rosen, the conflation of simple mechanisms with objectivity proves problematic when one recognizes that some systems cannot be adequately explained via reductionism. Remarking on the measurement problem in physics, Rosen notes that the resulting impredicativity precludes preservation of this traditional, mechanistic conception of objectivity:

> The problem here is that the very acquisition of data, the very cognition of phenomena… requires one to consider a *larger* system ("system + observer") and not to consider smaller ones, as reductionism (or context-independence, or objectivity) requires. This in turn creates a chicken-egg situation, an *impredicativity*; specifically, one must know the larger system to characterize the smaller, but one cannot know the larger until the smaller is characterized. (2000, p. 106, emphasis in original)

The problem is that one cannot fully define a system without reference to the larger system within which it is nested, and the larger system cannot be fully defined without reference to the original system in question. While Rosen is describing the measurement problem specifically, this problem of reciprocal reference does not only emerge in physical measurement; impredicativities are inherent to complex systems. A coherent and complete account of an organism, for example, requires reference to the interactions among its parts (e.g., metabolic and repair subsystems), but an account of these interacting parts requires reference to the larger-scale context (i.e., the organism). Following from this inability to conceptually remove systems from their contingent contexts, Rosen's (1958) relational biology places emphasis on the *relations* among these various scales of work that comprise a system. That is, rather than reducing phenomena to a set of fundamental parts and efficient causal processes, Rosen argues it is necessary to consider the organizational properties of a system. With this turn toward emphasizing organization, or multi-scale *relations*, scientific accounts of phenomena would no longer employ efficient causal stories and, instead, would incorporate explanations involving multi-scale constraints (i.e., how the various scales of contextual dynamics influence the activities that may be expressed by the system).

The use of these relation-laden explanations beckons one to recognize that phenomena are dependent on the contexts they are nested within (e.g., physical systems nested within the observational context or metabolic systems nested within the organismic context). Thus, impredicativities in scientific explanation problematize the traditional "perceiver-independent" and "context-independent" conception of objectivity because they indicate that phenomena cannot be defined independently of their contexts. For this reason, Rosen suggests that the popular insistence to reduce phenomena to simple mechanisms (and the conflation of objectivity with simple mechanisms) is driven by a "pervasive fear" of impredicativities. Nonetheless, he notes:

> The excluded process of finding out about a given system by putting it into a *larger* system (i.e., by creating a definite context for it), in the teeth of the impredicativities this generates, is perhaps the only exit from the boxes that reductionism alone, or mimesis alone, creates

for us. Biology already forces us out of these boxes, for life and mind are not to be found within them. (Rosen 2000, p. 107)

2 Wild Relationality

Both Silberstein and Chemero (2015) and Rosen (2000) challenge the distinction between subjective and objective properties by appealing to relational properties. Jordan and Day (2015) agree with this maneuver, and further ask to what extent our increasing explanatory reliance on relational properties might challenge our utilization of the notion of intrinsic properties at all. This idea has recently gained traction in the philosophy of science, as scholars have attempted to delineate what we mean when we distinguish intrinsic and relational properties. Mass, for example, is often treated as an intrinsic property because the mass of an object is thought to be independent of its context. Weight, on the other hand, is thought of as an extrinsic property because it is determined by how mass interacts with context. Despite our historical commitment to this distinction, Jammer (2009) proposes that inertial mass emerges as particles interact with the Higgs field, which is, "…a scalar field that "permeates all of space" and "endows particles with mass" (p. 162). Bauer (2011) argues such contextual dependence renders mass *externally grounded*. Jordan and Day (2015) assert that the notion of external grounding implies the mass of the particle is not independent of its context. As a result, the object's mass is a *relational, non-intrinsic* property. Consistent with this notion, Harré (1986) proposes the notion of *ultra-grounding*, which asserts that a property may be grounded by properties of reality as a whole.

Additional anti-intrinsic takes on the nature of properties can be found in Schaffer (2003) and Dehmelt (1989), who propose reality may be constituted of infinite levels of microstructure. Jordan and Day (2015) interpret this notion as implying that reality's lack of a final, fundamental level further implies there exist no final, non-relational, intrinsic properties. This assertion is consistent with the *Global Groundedness Thesis* (Prior et al. 1982), which claims all dispositions (i.e., properties) are grounded (i.e., externally grounded) rather than ungrounded (i.e., intrinsically grounded).

Clearly, philosophers of science are becoming increasingly skeptical of the existence of intrinsic properties. At the root of such concerns seems to be an upswing the in the belief that all properties are ultimately grounded in *context*. That is, no properties exist, as they do, independently of context. All properties are contextually bound. All properties emerge in a context. Though we might claim some properties, such as gravity, are *universal*, or *invariant* with respect to context, this does not mean they can exist, as they do, independent of context.

2.1 Implications of a Thoroughly-Relational Ontology

Given this inescapability of context, Jordan and Day (2015) consider the ontological possibilities that emerge when we conceptualize all properties as being relational. One such possibility is the ability to conceptualize all phenomena as constituting *embodiments of context*. From electrons, to atoms, to molecules, to organisms, to affordances, all phenomena are bound to context, emerge from context, and persist in context. As a result, they *all* constitute embodiments of their emergent contexts (Jordan and Ghin 2006). This is the essence of *wild relationality*. It is *wild* because its relationality is thoroughly contextual and contingent, at every level. Nothing intrinsic exists, save for the thoroughgoing relationality of *everything*.

The notion of *embodiments of context* lets us entertain the idea that all phenomena are naturally and necessarily *about* the contexts they embody. A cyclone that emerges over the Midwestern portion of the United States, for example, is naturally and necessarily *about* the global weather context in which it emerged. Jordan (2008) proposes such embodied aboutness renders all phenomena inherently meaningful, in that, they are naturally and necessarily *about* the contexts they embody.

To be sure, most cognitive scientists are not so liberal in their use of the concept *aboutness*, and, instead, utilize a more narrow definition relating to the aboutness inherent in the intentional nature of representations. Bickhard states, "The problem of representational content is a central aspect of the problem of intentionality—of how any system or agent can instantiate any sort of 'aboutness' relationship with its world." (1993, p. 1). Anderson and Rosenberg (2008) state, "How is one thing ever about another? To answer this question is usually to analyze this relation of aboutness—the intentionality of a representation—in terms of some other, presumably more basic relation" (p. 56). Intentionality, in these accounts, refers to the directedness of mental phenomena such as perceptions and thoughts. In other words, "…the mind's 'direction upon its objects', is what is distinctive of mental phenomena" (p. 1).

There are, of course, many, many different takes on the nature of intentionality and representation. Fodor (1981, 1987) proposes a *causal* theory in which a representation R is about T given it has a particular causal relation to T. Dretske (1981, 1986, 1988) proposes an *information-content* approach in which a representation is about the object that gave rise to the information the representation contains. Harman (1982, 1987) propose a Conceptual Role approach in which R is about E if R is used to make informed inferences about E. And Anderson and Rosenberg (2008) propose *guidance theory*, which asserts the intentionality of representations derives from the guidance they provide for actions. In all of these cases, the aboutness of the representation derives from its intentionality—from its directedness toward something beyond itself.

Clearly, there is a strong historical connection between the notions of intentional representation and the nature of the mental. As a result, it seems beyond the pale and somewhat bizarre to assert that embodiments of context are naturally and necessarily about the contexts they embody. For if this is the case, then representation, aboutness and intentionality are ubiquitous. If all properties are wildly relational, then all of reality in necessarily *about* all of reality. In short, if reality is constituted of relational properties, it is constituted of *aboutness*. It is constituted of *meaning*.

Such an approach to aboutness in no way negates causal, information-content, conceptual role, or guidance theory accounts of representation. To the contrary, by conceptualizing all of reality as being constituted of embodied aboutness, the notion of wild relationality provides sound ontological footing for the types of aboutness reflected in the afore-mentioned accounts. Said another way, wild relationality provides a means of solving the grounding problem (Harnad 1990). It does so by shifting our ontological focus. Instead of trying to determine how physical symbol systems are able to carry representational content, the notion of wild relationality argues all properties are inherently relational, embodiments of context. In short, *embodied context* replaces the concepts "physical" and "mental" (Jordan and Day 2015). As a result, the types of situations addressed by more narrowly-focused accounts of representation can be seen as being naturally and necessarily about, and what differs between them is the contexts in which their abountness manifests itself.

2.2 Embodiments of Context and Wild Systems

The notions of *wild relationality* and *embodiments of context* stem from Wild Systems Theory (WST—Jordan 2013), a scientifically-grounded approach to describing organisms that focuses on their status as thoroughly relational embodiments of context, versus physical systems having objective and/or subjective properties. Describing organisms in this manner leads to the issue of whether or not cognition is a unique property in the universe. We will come back to this question after thoroughly reviewing what WST says about organisms as embodiments of context.

Following the lead of Schrödinger (1944), Odum (1988), Maturana and Varela (1980), Rosen (1958), and Kauffman (1995), WST conceptualizes living systems as multi-scale nestings of self-sustaining, energy-transformation systems. That is, organisms are far-from-equilibrium systems whose ability to intake, transform, and dissipate energy actually keeps them far from thermodynamic equilibrium. Given this emphasis on open, dissipative systems, WST is clearly consistent with contemporary science. Instead of describing organisms as being fundamentally physical, chemical, or biological however, WST focuses on the dynamics of energy transformation. As we shall see, doing so prevents WST from having to rely on concepts such as physical, mental, and/or information.

WST and representation. In WST, the message *is* the medium, in that, the *stabilities* referred to via the concepts *neurons*, *neural networks*, *global brain states*, and the *body*, all constitute embodiments of context that emerge dynamically and *contextually* (Atmanspacher 2007; Bishop and Atmanspacher 2006) via energy transformation processes that produce products (e.g., catalysts) that feedback into and, ultimately *sustain* the stability-producing energy transformations. In short, these stabilities *are* the self-sustaining, dynamic work (i.e., energy transformations) that constitutes them.

Jordan and Vinson (2012) describe how the notion of self-sustaining work has been discovered by different scholars, in different disciplines, at different points in history.

> At the chemical level, self-sustaining work has been referred to as autocatalysis (Kauffman 1995), the idea being that a self-sustaining chemical system is one in which reactions produce either their own catalysts or catalysts for some other reaction in the system. At the biological level, self-sustaining work has been referred to as autopoiesis (Maturana and Varela 1980), again, the idea being that a single cell constitutes a multi-scale system of work in which lower-scale chemical processes give rise to the larger biological whole of the cell which, in turn, provides a context in which the lower-scale work sustains itself and the whole it gives rise to (Jordan and Ghin 2006). Hebb (1949) referred to the self-sustaining nature of neural networks as the 'cell assembly', the idea being that neurons that fire together wire together. Jordan and Heidenreich (2010) recently cast this idea in terms of self-sustaining work by examining data that indicate the generation of action potentials increases nuclear transcription processes in neurons which, in turn, fosters synapse formation. At the behavioural level, Skinner (1976) referred to the self-sustaining nature of behaviour as operant conditioning, the idea being that behaviours sustain themselves in one's behavioural repertoire as a function of the consequences they generate. Streeck and Jordan (2009) recently described communication as a dynamical self-sustaining system in which multi-scale events such as postural alignment, gesture, gaze, and speech produce outcomes that sustain an ongoing interaction. And finally, Odum (1988) and Vandervert (1995) used the notion of self-sustaining work to refer to ecologies in general. (p. 235)

WST brings together these various takes on the notion of self-sustaining work, and combines them in a way that reveals the homological unity of the energy-transformation dynamics that transcend the phyla, from the autocatalytic dynamics of certain chemical systems, to the self-sustaining dynamics of human interaction.

At every level of scale, self-sustaining systems establish and maintain their status as systems because their constitutive work gives rise to and sustains a permeable, system-context border. In the case of a single neuron, the border is a lipid bilayer. In the case of a society working to sustain a certain pattern of relations with context, the *border* involves multiple, nested, time-scales of context:

> Like any ecosystem, the cultural-cognitive ecosystem can be seen as a constraint satisfaction system that settles into a subset of possible configurations of elements. It is a dynamical system in which certain configuration of elements (what we know as stable practices) emerge (self-assemble) preferentially. In this perspective, constraints exist in many places and interact with one another through a variety of mechanisms of constraint satisfaction. Some of these are neural mechanisms; others are implemented in material tools; and still others are emergent in social processes of collective intelligence, the development of conventions, for example. (Hutchins 2014, p. 46)

Given the borders generated by self-sustaining systems are necessarily porous, at all levels of scale, they are also necessarily *about* the context in which the system sustains itself. In the case of a single neuron, the lipid bilayer is constituted of proteins that allow certain chemicals to enter the neuron, while simultaneously keeping others out. In essence, the lipid bilayer constitutes an *embodiment* of the constraints the system needs to address in order to persist as a system in its context. From this perspective, the border itself is inherently relational. Said another way, the border is naturally and necessarily about the context it embodies. In short, it "represents" its context. As a result, the systems does not have to perceive, detect, capture or process *information* in order to have meaning (Jordan et al. 2017). As an embodiment of context, the system is constituted of embodied aboutness because it is a thoroughly relational system. In short, it "represents" its context and is therefore inherently *meaningful*.

By conceptualizing self-sustaining systems as *embodied context*, WST does not have to address such systems in terms of objective or subjective properties. Instead, WST describes organism-environment coordinations in terms of *mutual modulation*. That is, self-sustaining systems necessarily modulate the contexts in which they are embedded, while the embedding context simultaneously modulates self-sustaining systems. At no point in such interactions is there a subjective-objective gap to be crossed. As a result, there is no need for the organism to generate representations of its environment. Every aspect of the organism is a "re-presentation" to the world of the phylogenetic, cultural, social, and ontogentic contexts in which the system emerged and sustains itself.

Scaling up the representational content of wild systems. Given Jordan and Vinson's (2012) analysis of the multiple scales of self-sustaining systems that constitute a human being, we can conceptualize ourselves as embodiments of multiple, nested, time-scales of context, in which the time scales range from the phylogenetic, to the cultural, the social, and the ontogenetic. Our neuromuscular architecture, for example, can be seen as a multi-scale embodiment of the constraints that have to be addressed to propel a mass, as a whole, through a gravity field. Our ability to write this paper reflects the interaction of multiple, nested time scales of self-sustaining work, including each of our individual life trajectories that led us to this point, as well as the multiple meetings we held to discuss the paper (Lemke 2000). As an example of the interaction of these nested scales of self-sustaining work, Streeck and Jordan (2009) state the following:

> While different time-scale of behavior are all implicated in structuring participants' interaction with one another, it is difficult to assign meaning to them within a traditional information-processing framework: They are meaningful on in terms of the ways in which they constrain and contextualize the immediate interactional context within which they are occur (e.g., the sequence of communicative acts) and of how features of immediate contexts in turn constrain and contextualize higher-level or remote contexts (e.g., the social relationships that endure beyond the single encounter). (p. 453)

WST and the intentionality of representations. The Stanford Encyclopedia of Philosophy (Siewert 2017) states:

> Intentionality...has to do with the directedness, aboutness, or reference of mental states—the fact that, for example, you think of or about something. Intentionality includes, and is sometimes seen as equivalent to, what is called "mental representation".

In contrast, Jordan and Ghin (2006) describe self-sustaining systems as a form of end-directedness in which the "end" is implicit in the dynamics of the system's self-sustaining work.

> By "end-directed" we do not mean to imply such systems have "goals" that guide their "behavior." Rather, we mean that the micro-macro synergies that constitute self-sustaining

> systems are inherently such that the work of the system is able to produce products that keep energy coming into the system. (p. 40)

Given that WST conceptualizes organisms as a multi-scale nesting of self-sustaining systems, Jordan and Heidenreich (2010) argue that every level of self-sustaining work constitutes a level of end-directedness—of intentionality. Given every level of work is both intentional and "represents", harboring intentional content is not a property belonging solely to brains. Rather, brain dynamics are capable of endogenous and exogenous modulation on time scales much faster than those of, say, the body as a whole. Thus, according to WST, while both levels of self-sustaining work are necessarily end-directed toward sustainment (i.e., intentional) and constitute embodiments of context (i.e., constitute representations), it is the ones taking place in brain dynamics that we have been led to describe as being cognitive, and therefore, of a different kind than other levels of self-sustaining work.

2.3 *Wild Systems, Embodied Contexts, and Cognition*

Given our thoroughly relational account of wild systems (i.e., organisms) as multi-scale, self-sustaining embodiments of context, we are not sure what is gained by working to clearly delineate the boundaries of cognition. WST reveals the homological unity of self-sustaining energy-transformation dynamics that transcend the phyla, from the simplest living organism to the most complex. Given such systems are inherently intentional (i.e., end-directed) and representational (i.e., they constitute embodied context), it is not clear how one level of self-sustaining work such as *thought* truly differs from other levels. To be sure, *thoughts* allow us to sustain relationships with events that exist at much larger time scales than those involved in simply moving our bodies. But in the end, the self-sustaining dynamics are homologous. In addition, WST's conceptualization of organisms as embodiments of context reveals such systems cannot exist outside of contexts that afford their existence. Said another way, exogenous factors constitute a necessary aspect of self-sustaining work. As a result, it makes no sense to ask if exogenous factors are constitutive of self-sustaining work. The answer is obviously 'yes'.

In all honesty, it may be the case that the search for the 'mark of the cognitive' stems from our commitment to the distinction between objective and subjective properties. For once we believe there is a subjectivity-objectivity gap, we feel compelled to posit isolated, internally located *minds* that need to *make contact* with external reality. Representations and computations are then proposed as mechanisms that allow such contact.

By conceptualizing living systems as embodiments of context, versus as physical systems that process and/or detect information, WST evades the *inherently meaningless* view of reality entailed in contemporary, naturalism, "...the harshness of naturalist metaphysics exactly consists in the point that nothing has intrinsic value, because any possible or actual fact is only normative relative to a certain organism, biological population, self-sustaining robot, or other such entity" (Metzinger 2017, p. 18). As a result, WST is not concerned with discovering the mark of the cognitive. Phenomena referred to via concepts such as *reading*, *problem-solving*, *theory-of-mind*, and *mental rotation*, do exist. But calling them *cognitive* stems from a commitment to a subjective-objective gap WST was built to avoid.

3 Conclusions

Within the thoroughly relational framework posited by WST, the skin *is* a border, but not an epistemic one. It is a self-sustaining border that is generated and sustained by the energy transformation dynamics that constitute it. As a result, knowledge is not something we have to obtain in order to be about the contexts we embody and are embedded within. Knowledge *is* the energy-transformation systems that constitute us, because such systems are naturally and necessarily about the contexts they embody. What we call cognition, then, is a phylogentically, scaled-up recursion on the inherently meaningful relationality of which all self-sustaining systems are constituted. And if some aspects of our sustained relationality necessarily lie outside the border of our skin, so be it.

References

Adams F, Aizawa K (2001) The bounds of cognition. Philos Psychol 14(1):43–64

Anderson ML, Rosenberg G (2008) Content and action: the guidance theory of representation. J Mind Behav 29(1, 2):55–86

Atmanspacher H (2007) Contextual emergence from physics to cognitive neuroscience. J Conscious Stud 14(1–2):18–36

Bauer W (2011) An argument for the extrinsic grounding of mass. Erkenntnis 74(1):81–99. https://doi.org/10.1007/s10670-010-9269-4

Bickhard MH (1993) Representational content in humans and machines. J Exp Theor Artif Intell 5:285–333

Bishop RC, Atmanspacher H (2006) Contextual emergence in the description of properties. Found Phys 36(12):1753–1777

Chalmers DJ (1996) The conscious mind: in search of a fundamental theory. Oxford University Press

Chemero A (2011) Radical embodied cognitive science. MIT Press

Dehmelt H (1989) Triton, electron, cosmon: an infinite regression? Proc Natl Acad Sci 86(22):8618–8619. https://doi.org/10.1073/pnas.88.4.1590a

Dretske F (1981) Knowledge and the flow of information. The MIT Press, Cambridge, Massachusetts

Dretske F (1986) Misrepresentation. In: Bogdan R (ed) Belief: form, content, and function. Oxford University Press, New York, pp 17–36

Dretske F (1988) Explaining behavior. The MIT Press, Cambridge, Massachusetts

Fodor J (1981) Representations: philosophical essays on the foundations of cognitive science. The MIT Press, Cambridge, Massachusetts

Fodor J (1987) Psychosemantics. The MIT Press, Cambridge, Massachusetts

Harman G (1982) Conceptual role semantics. Notre Dame J Form Log 23:242–256

Harman G (1987) (Nonsolipsistic) conceptual role semantics. In: LePore E (ed) Semantics of natural language. Academic Press, New York, pp 55–81

Harnad S (1990) The symbol grounding problem. Physica D 42(1–3):335–346

Harré R (1986) Varieties of realism: a rationale for the natural sciences. Blackwell, Oxford

Hebb DO (1949) The organization of behavior: a neuropsychological theory. Wiley, New York, NY

Hutchins E (2014) The cultural ecosystem of human cognition. Philos Psychol 27(1):34–49

Jammer M (2009) Concepts of mass in contemporary physics and philosophy. Princeton University Press, Princeton, NJ

Jordan JS (2008) Wild agency: nested intentionalities in cognitive neuroscience and archaeology. Philos Trans R Soc B: Biol Sci 363(1499):1981–1991

Jordan JS (2013) The wild ways of conscious will: what we do, how we do it, and why it has meaning. Front Psychol 4:574

Jordan JS, Cialdella VT, Dayer A, Langley MD, Stillman Z (2017) Wild bodies don't need to perceive, detect, capture, or create meaning: they are meaning. Front Psychol 8:1149

Jordan JS, Day B (2015) Wild systems theory as a 21st century coherence framework for cognitive science. In: Metzinger T, Windt JM (eds) Open MIND: 21. MIND Group, Frankfurt am Main. https://doi.org/10.15502/9783958570191

Jordan JS, Ghin M (2006) (Proto-) consciousness as a contextually emergent property of self-sustaining systems. Mind Matter 4(1):45–68

Jordan JS, Heidenreich BA (2010) The intentional nature of self-sustaining systems. Mind Matter 8(1):45–62

Jordan JS, Vinson D (2012) After nature: on bodies, consciousness, and causality. J Conscious Stud 19(5–6):229–250

Kauffman S (1995) At home in the universe. Oxford University Press, New York, NY

Lemke J (2000) Across the scales of time: artifacts, activities, and meaning in eco-social systems. Mind Cult Act 7(4):273–290

Maturana HR, Varela FJ (1980) Problems in the neurophysiology of cognition. In: Autopoiesis and cognition. Springer, Dordrecht, pp 41–47

Metzinger T (2017) The problem of mental action: predictive control without sensory sheets. Johannes Gutenberg-Universität Mainz

Odum HT (1988) Self-organization, transformity, and information. Science 242(4882):1132–1139. https://doi.org/10.1126/science.242.4882.1132

Prior E, Pargetter R, Jackson F (1982) Three theses about dispositions. Am Philos Q 19(3):251–257

Raja V, Biener Z, Chemero A (2017) From Kepler to Gibson. Ecol Psychol 29(2):146–160

Rosen R (1958) A relational theory of biological systems. Bull Math Biophys 20:245–260. https://doi.org/10.1007/BF02478302

Rosen R (2000) Essays on life itself. Columbia University Press, New York

Schaffer J (2003) Is there a fundamental level? Noûs 37(3):498–517. https://doi.org/10.1111/1468-0068.00448

Schrödinger E (1944) What is life? The physical aspect of the living cell and mind. Cambridge University Press, Cambridge

Siewert C (2017) Consciousness and intentionality. In E. Zala's (Ed.), The Stanford Encyclopedia of Philosophy. https://plato.stanford.edu/archives/spr2017/entries/consciousness-intentionality

Silberstein M, Chemero A (2011) Dynamics, agency and intentional action. Humana Mente 15:1–19

Silberstein M, Chemero A (2015) Extending neutral monism to the hard problem. J Conscious Stud 22(3–4):181–194

Skinner BF (1976) About behaviorism. Vintage Books, New York, NY

Smith LB, Thelen E (2003) Development as a dynamic system. Trends Cognit Sci 7(8):343–348

Streeck J, Jordan JS (2009) Communication as a dynamical self-sustaining system: the importance of time-scales and nested context. Commun Theory 19(4):445–464

Vandervert L (1995) Chaos theory and the evolution of consciousness and mind: a thermodynamic-holographic resolution to the mind–body problem. New Ideas Psychol 13(2):107–127. https://doi.org/10.1016/0732-118x(94)00047-7

Fictionalists Disregard the Dynamic Nature of Scientific Models

Lorenzo Magnani

Abstract In the current epistemological debate scientific models are not only considered as useful devices for explaining facts or discovering new entities, laws, and theories, but also rubricated under various new labels: from the classical ones, as abstract entities and idealizations, to the more recent, as fictions, surrogates, credible worlds, missing systems, make-believe, parables, functional, epistemic actions, revealing capacities. An influential article by John Woods, entitled "Against fictionalism" (Woods 2013), usefully provides a rich argumentation concerning the puzzling problems created by the use of the concept of fiction in philosophy, epistemology, and logic, Woods himself further deepened in the recent book *Truth in Fiction: Rethinking Its Logic* (Woods 2018). By limiting my treatment to the case of models in science, I would like to offer an additional support to this perspective, emphasizing the unsatisfactory character of this intellectual recent trend, and the uselessness of the concept of fiction in illustrating the scientific enterprise. I will contend that it is misleading to analyze models in science by disregarding the dynamic aspects: scientific models in a static perspective (for example when inserted in a textbook) certainly appear fictional to the epistemologist, but their fictional character disappears if a dynamic perspective is adopted. The article also sketches the role of models in science taking advantage of the concept of "epistemic warfare", which sees scientific enterprise as a complicated struggle for rational knowledge in which it is crucial to distinguish epistemic (for example scientific models) from non epistemic (for example fictions, falsities, propaganda) weapons. A reference to the usefulness of Feyerabend's counterinduction in criticizing the role of resemblance in model-based cognition is also provided, to further corroborate the thesis indicated by the article title.

Keywords Fictions · Fictionalism · Scientific models · Model-based reasoning · Dynamics of scientific knowledge · Static view of science

L. Magnani (✉)
Department of Humanities, Philosophy Section and Computational Philosophy Laboratory, University of Pavia, Pavia, Italy
e-mail: lmagnani@unipv.it

T. Bertolotti (ed.), *Cognition in 3E: Emergent, Embodied, Extended*, Studies in Applied Philosophy, Epistemology and Rational Ethics 56,
https://doi.org/10.1007/978-3-030-46339-7_5

1 Are Scientific Models Fictions?

An influential article by John Woods, entitled "Against fictionalism" (Woods 2013), usefully provides a rich argumentation concerning the puzzling problems created by the use of the concept of fiction in philosophy, epistemology, and logic, Woods himself further deepened in the recent book *Truth in Fiction: Rethinking Its Logic* (Woods 2018). By limiting my treatment to the case of models in science, I would like to offer an additional support to this perspective, emphasizing the unsatisfactory character of this intellectual recent trend, and the uselessness of the concept of fiction in illustrating scientific enterprise. The article also sketches the role of modeling activity in science taking advantage of the concept of "epistemic warfare", which sees scientific enterprise as a complicated struggle for rational knowledge in which it is crucial to distinguish epistemic (for example scientific models) from non epistemic (for example fictions, falsities, propaganda) weapons. However, my main critique will be substantiated by the observation that it is easier to legitimate the role of models as fictions adopting a static view of science, but much more difficult if we adopt a perspective that stresses the dynamic of scientific reasoning.

Scientific models are now not only considered as useful ways for explaining facts or discovering new entities, laws, and theories, but are also rubricated under various new labels: from the classical ones, abstract entities (Giere 1988, 2007, 2009) and idealizations (Mizrahi 2011; Portides 2007; Weisberg 2007), to the more recent, fictions (Contessa 2010; Fine 2009; Frigg 2010a, b, c; Godfrey-Smith 2006, 2009; Suárez 2009, 2010; Woods and Rosales 2010a, b; Woods 2010), surrogates (Contessa 2007), credible worlds (Kuorikoski and Lehtinen 2009; Sugden 2000, 2009), missing systems (Mäki 2009; Thomson-Jones 2010), as make-believe (Frigg 2010a, b, c; Toon 2010), parables (Cartwright 2009b), as functional (Chakravartty 2010), as epistemic actions (Magnani 2004a, b), as revealing capacities (Cartwright 2009a). This proliferation of explanatory metaphors is amazing, if we consider the huge quantity of knowledge on scientific models that had already been produced both in epistemology and in cognitive science. Some of the authors mentioned above are also engaged in a controversy about the legitimacy especially of speaking of fictions in the case of scientific models.

Even if the above studies related to fictionalism have increased knowledge about some aspects of the role of models in science, I am convinced that sometimes they have also generated some philosophical confusion and it seems to me correct (following the suggestion embedded in the title of a recent paper) "to keep quiet on the ontology of models" (French 2010), and also to adopt a more skeptical theoretical attitude. I think that, for example, models can be considered fictions or surrogates, but this just coincides with a common sense view, which appears to be philosophically empty or, at least, delusory. Models are used in a variety of ways in scientific practice, they can also work as mediators between theory and experiment (Portides 2007), as pedagogical devices, for testing hypotheses, or for explanatory functions (Bokulich 2011), but these last roles of models in science are relatively well-known and weakly disputed in the epistemological literature. In this paper I will concentrate on scientific models— seen in a perspective which takes advantage of a distributed cognition framework—in creative abductive cognitive processes, which I still consider the central problem of current epistemological research (Hintikka 1998).

I think that models, both in scientific reasoning and in human perception, are neither mere fictions, simple surrogates or make-believe, nor they are unproblematic idealizations; in particular, models are never *abstract*, contrarily to the received view. Indeed, within science the adopted models are certainly constructed on the basis of multiple constraints relating to the abstract laws, principles, and concepts, when clearly available at a certain moment of the development of a scientific discipline. At the same time we have to immediately stress that the same models are always *distributed* material entities, either when we are dealing with concrete diagrams or physical and computational models, or when we face human "mental models", which at the end are indeed particular, unrepeatable, and ever-changing configurations and transformations of neural networks and chemical distributions at the level of human brains. In this perspective we can say that models are "abstract" only in a Pickwickian sense, that is as "mental models", shared to different extents by groups of scientists, depending on the type of research community at stake. This cognitive perspective can therefore help us in getting rid of the ambiguities sparked by the notion of abstractness of models.

In a previous work I have also enriched my critique of fictionalism outlining the first features of my own approach to the role of scientific models in terms of what I call "epistemic warfare" (cf. (Magnani 2017, chapter four)), which sees scientific enterprise as a complicated struggle for rational knowledge in which it is crucial to distinguish epistemic (for example scientific models) from non epistemic (for example fictions, falsities, propaganda, etc.) weapons. The characteristic feature of *epistemic* weapons is that they are value-directed to the aim of promoting the attainment of scientific truth, for example through predictive and empirical accuracy, simplicity, testability, consistency, etc.[1] Indeed we have to consider scientific enterprise a complicated epistemic warfare, so that we could plausibly expect to find fictions in this struggle for rational knowledge. Are not fictions typical of any struggle which characterizes the conflict of human coalitions of any kind? During the Seventies of the last century Feyerabend (1975) clearly stressed how, despite their eventual success, the scientist's claims are often far from being evenly proved, and accompanied by "propaganda [and] psychological tricks in addition to whatever intellectual reasons he has to offer" (p. 65), like in the case of Galileo. Galileo's discussions of real experiments—in the *Dialogo* but also in the *Discorsi*—become rhetorical, to confound the opponents and persuade the readers, and also to fulfil didactic needs, as contended by Naylor (1976).

These tricks are very useful and efficient, but one thing is the *epistemic* role of reasons scientist takes advantage of, such the scientific models I will illustrate in this paper, which for example directly govern the path to provide a new intelligibility of the target systems at hand; another thing is the *extra-epistemic* role of propaganda

[1]In this perspective I basically agree with the distinction between epistemic and non-epistemic values as limpidly depicted in (Steel 2010).

and rhetoric, which only plays a mere—positive or negative—ancillary role in the epistemic warfare. So to say, these last aspects support scientific reasoning providing non-epistemic weapons able for example to persuade other scientists belonging to a rival "coalition" or to build and strengthen the coalition in question, which supports a specific research program, for example to get funds.

I am neither denying that models as idealizations and abstractions are a pervasive and permanent feature of science, nor that models, which are produced with the aim of finding the consequences of theories—often very smart and creative—are very important. I just stress that the "fundamental" role played by models in science is the one we find in the *dynamical* perspective of the the core conceptual discovery processes, and that these kinds of models cannot be indicated as fictional at all, because they are *constitutive* of new scientific frameworks and new empirical domains. In this last sense the capacity of scientific models to constitute new empirical domains and so new *knowability* is ideally related to the emphasis that epistemology, in the last century, put on the theory-ladenness of scientific facts (Hanson, Popper, Lakatos, Kuhn): in this light, the formulation of observation statements presupposes significant knowledge, and the search for new observability in science is guided by scientific modeling.

Suárez (2009) provides some case studies, especially from astrophysics and concerning quantum model of measurement, emphasizing the inferential function of the supposed to be "fictional" assumptions in models: I deem this function to be ancillary in science, even if often highly innovative. Speaking of the Thomson's plum pudding model Suárez maintains that, basically "The model served an essential pragmatic purpose in generating quick and expedient inference at the theoretical level, and then in turn from the theoretical to the experimental level. It articulated a space of reasons, a background of assumptions against which the participants in the debates could sustain their arguments for and against these three hypotheses" (p. 163). In these cases the fact that various assumptions of the models are empirically false is pretty clear and so is the "improvement in the expediency of the inferences that can be drawn from the models to the observable quantities" (p. 165)[2]: the problem is that in these cases models, however, are not fictions—at least in the minimal unequivocal sense of the word as it is adopted in the literary/narrative frameworks—but just the usual idealizations or abstractions, already well-known and well studied, as devices, stratagems, and strategies that lead to efficient results and that are not discarded just because they are not fake chances from the perspective of scientific rationality.[3] Two consequences derive:

---

[2]It has to be added that Suárez does not conflate scientific modeling with literary fictionalizing and distinguishes scientific fictions from other kinds of fictions—the scientific ones are constrained by both the logic of inference and, in particular, the requirement to fit in with the empirical domain (Suárez 2009, 2010)—in the framework of an envisaged compatibility of "scientific fiction" with realism. This epistemological acknowledgment is not often present in other stronger followers of fictionalism.

[3]I discussed the role of chance-seeking in scientific discovery in (Magnani 2007). For a broader discussion on the role of luck and chance-seeking in abductive cognition see also (Bardone 2011), and (Bardone 2012).

- the role of models as "expediency of the inferences" in peripheral aspects of scientific research, well-known from centuries in science, does not have to be confused with the *constitutive* role of modeling in the central creative processes, when new conceptually revolutionary perspectives are advanced: in this case the *dynamic* role of models in scientific cognition dominates;
- models are—so to say—just models that idealize and/or abstract, but these last two aspects have to be strictly criticized in the light of recent epistemologico/cognitive literature as related to special kinds of epistemic actions, as I have illustrated in (Magnani 2012): abstractness and ideality cannot be solely related to empirical inadequacy and/or to theoretical incoherence (Suárez 2009, p. 168), in a *static* view of the scientific enterprise.

In sum, I will illustrate that there is no need of reframing—in the new complicated and intellectualistic lexicon of fictions (and of the related metaphors)—what is already well-known thanks to the tradition of philosophy of science. We have to remorselessly come back to the spirit of Newton's famous motto "hypotheses non fingo", which has characterized for centuries the spirit of modern science: "I have not as yet been able to discover the reason for these properties of gravity from phenomena, and I do not feign hypotheses. For whatever is not deduced from the phenomena must be called a hypothesis; and hypotheses, whether metaphysical or physical, or based on occult qualities, or mechanical, have no place in experimental philosophy. In this philosophy particular propositions are inferred from the phenomena, and afterwards rendered general by induction" (Newton 1999, p. 493).

2 Are the In-Vitro Model or a Geometrical Diagram Fictions? Dynamic Versus Static View of Scientific Models

Peirce, speaking of the model-based aspects of deductive reasoning, hypothesized there is an "experimenting upon this image [the external model/diagram] in the imagination", so showing how human geometrical imagination is always triggered by a kind of prosthesis, the external model as an "external imagination". Analogously, taking advantage of a fictional view on models and of the pretence theory Frigg (2010c, p. 266 ff.) interestingly sees imagination as an authorized intersubjective game of make-believe sanctioned by the "prop" (an object, for example material models, movies, paintings, plays, etc.) and its rules of generation. This theory also works as a metaphor of abductive processes (that is cognitive processes that lead to hypotheses), in terms of some concepts taken from the theory of literary and artistic fictions. Again, I think that it is neither necessary to adopt a fictionalist view in the case of science, nor the pretence theory adds something relevant to the issue. For example, scientists that build models in a lab do not pretend anything and are not engaged in the relative make-believe process, if not in the trivial sense that almost every human intersubjective interplay can be seen as such. For instance, in-vitro networks of cultured neurons or the Peircean Euclidean diagram used by the ancient Greek geometers are just the opposite of a mere fiction or of a generic make-believe interplay, they are instead more or less mimetic (possibly creative) "distributed" external models which are expected to provide reliable information about the target system. They aim at discovering some new representations about the neurons in the first case and about the pure concepts of geometry in the second.

The reason of my skepticism can be illustrated taking advantage of some theses derived from classical Kantian philosophy and Thom's mathematical semiophysics. Immanuel Kant was clearly aware of the interplay between internal and external models, exemplified in the case of a formal science like mathematics. In his transcendental terms, Kant says that in geometrical construction "[…] I must not restrict my attention to what I am actually thinking in my concept of a triangle (this is nothing more than the mere definition); I must pass beyond it to properties which are not contained in this concept, but yet belong to it" (Kant 1929, A718-B746, p. 580). Hence, for Kant models in science (in this case, of geometry) are first of all *constructions* that go beyond what the researcher simply "thinks". The same situations can be now illustrated as a case of what I called manipulative abduction (Magnani 2001, 2009), that is is a kind of, usually model-based, abduction that exploits external models endowed with delegated (and often implicit) cognitive roles and attributes: 1. The model is external and the strategy that organizes the manipulations is unknown a priori. 2. The result achieved is new (if we, for instance in this geometrical case, refer to the constructions of the first creators of geometry), and adds properties not contained before in the concept (the Kantian to "pass beyond" or "advance beyond" the given concept (Kant 1929, A154-B194, p. 192)).[4]

Iconicity of models is central for Peirce, who analogously to Kant, maintains that "[…] philosophical reasoning is reasoning with words; while theorematic reasoning, or mathematical reasoning is reasoning with specially constructed schemata" (Peirce 1931–1958, 4.233); moreover, he uses diagrammatic and schematic as synonyms, thus relating his considerations to the Kantian tradition where schemata mediate between intellect and phenomena. Schematism, a fruit of the imagination is, according to Kant, "[…] an art concealed in the depths of the human soul, whose real modes of activity nature is hardly likely ever to allow us to discover, and to have open to our gaze" (Kant 1929, A141-B181, p. 183).

Now we have at our disposal, thanks to epistemology and cognitive science, a lot of knowledge about the cognitive processes which correspond to Kantian schematism[5]. The following is the famous related passage in the *Critique of Pure Reason* ("Transcendental Doctrine of Method"):

> Suppose a philosopher be given the concept of a triangle and he be left to find out, in his own way, what relation the sum of its angles bears to a right angle. He has nothing but the concept of a figure enclosed by three straight lines, and possessing three angles. However

[4]Of course in the case we are using diagrams to demonstrate already known theorems (for instance in didactic settings), the strategy of manipulations is often already available and the result is not new.

[5]On models as epistemic mediators in mathematics cf. (Boumans 2012).

> long he meditates on this concept, he will never produce anything new. He can analyse and clarify the concept of a straight line or of an angle or of the number three, but he can never arrive at any properties not already contained in these concepts. Now let the geometrician take up these questions. He at once begins by constructing a triangle. Since he knows that the sum of two right angles is exactly equal to the sum of all the adjacent angles which can be constructed from a single point on a straight line, he prolongs one side of his triangle and obtains two adjacent angles, which together are equal to two right angles. He then divides the external angle by drawing a line parallel to the opposite side of the triangle, and observes that he has thus obtained an external adjacent angle which is equal to an internal angle—and so on. In this fashion, through a chain of inferences guided throughout by intuition, he arrives at a fully evident and universally valid solution of the problem (Kant 1929, A716-B744, pp. 578–579).

Here "intuition" is the Kantian word that expresses our present reference to what we call "external—scientific, geometrical in this case—model".

We can depict the situation of the philosopher described by Kant at the beginning of the previous passage taking advantage of some ideas coming from Thom's catastrophe theory. As a human being who is not able to produce anything new relating to the angles of the triangle, the philosopher experiences a feeling of frustration (just like the Kölher's monkey which cannot keep the banana out of reach). The bad affective experience "deforms" the organism's regulatory structure by complicating it and the cognitive process stops altogether. The geometer instead "at once constructs the triangle" [the scientist constructs the model] that is, he makes an external representation of a triangle and acts on it with suitable manipulations. Thom thinks that this action is triggered by a "sleeping phase" generated by possible previous frustrations which then change the cognitive status of the geometer's available and correct internal idea of triangle (like the philosopher, he "has nothing but the concept of a figure enclosed by three straight lines, and possessing three angles", but his action is triggered by a sleeping phase). Here the idea of the triangle is no longer the occasion for "meditation", "analysis" and "clarification" of the "concepts" at play, like in the case of the "philosopher". Here the inner concept of triangle—symbolized as insufficient—is amplified and transformed thanks to the sleeping phase (a kind of Kantian imagination active through schematization) in a prosthetic triangle to be put outside, in some external support. The instrument (here an external diagram) becomes the extension of an organ:

> What is strictly speaking the end—to take the banana [in our case, to find the sum of the internal angles of a triangle]—must be set aside in order to concentrate on the means of getting there. Thus the problem arises, a sort of vague notion altogether suggested by the state of privation. [...] As a science, heuristics does not exist. There is only one possible explanation: the affective trauma of privation leads to a folding of the regulation figure. But if it is to be stabilized, there must be some *exterior form* [italics added] to hold on to. So this anchorage problem remains whole and the above considerations provide no answer as to why the folding is stabilized in certain animals or certain human beings whilst in others (the majority of cases, needless to say!) it fails (Thom 1988, pp. 63–64).[6]

[6] A full analysis of the Kölher's chimpanzee getting hold of a stick to knock a banana hanging out of reach in terms of the mathematical models of the perception and the capture catastrophes is given in (Thom 1988, pp. 62–64). On the role of emotions, for example frustration, in scientific discovery cf. (Thagard 2002).

3 The Epistemic Danger of Disregarding the Dynamic Aspects of the Scientific Enterprise

Taking advantage of Thom's considerations, we can clearly see that the constructed external scientific model in the case of creative processes is exactly the opposite both of a fiction and of a generic process of make-believe (neither is a mere surrogate (Contessa 2007) or a bare credible world (Sugden 2000, 2009). It is instead a *regulatory* tool *stabilized* in "some exterior form", a kind of a reliable anchorage, not intentionally established as fiction, as a romance writer could intentionally do, assessing the fictional character of Anna Karenina. In the epistemological fictionalism about models the use of the label "fiction" is usually legitimated by the fact that there are no empirical systems corresponding for example to the ideal pendulum (and its equation).

Unfortunately the label sets up a paradox we can clearly see taking advantage of the case of scientific models seen as "missing systems", another new metaphor that echoes the fictional one— indeed the description of a missing system might be a fiction. Thomson-Jones (2010) emphasizes that science is full of "descriptions of missing systems", that at the end are thought as *abstract models*.[7] Further, Mäki (2009) usefully acknowledges that scientific models are "pragmatically and ontologically constrained representations", and further complicates the missing systems framework adding a supplementary metaphoric conceptual apparatus: missing systems are also "surrogate" systems expressed as credible worlds, as models. Similar arguments are advanced by Godfrey-Smith (2009, p. 114): "To say that talk of model systems is a psychologically exotic way of investigating conditionals (and the like) is not itself to solve the problem. It is natural to think that the useable output we get from modeling is generally a conditional—a claim that if such-and such a configuration existed, it would behave in a certain way. The configurations in question, however, are usually known *not* to exist, so the problem of explaining the empirical usefulness of this kind of knowledge reappears".

I contend that, at least in the dynamics of a discovery cognitive process, the missing system (Thomson-Jones) is not, paradoxically, the one represented by the "model", but instead the target system itself, still more or less largely unknown and un-schematized, which will instead appear as "known" in a new way only after the acceptation of the research process results, thus admitted into the theory T and considered worth to *staying* in T thereafter. The same can be said of models as configurations (Godfrey-Smith), which certainly are conditional, but at the same time do not have to be considered as "known *not* to exist", in Godfrey-Smith's sense, because simply in the moment in which a scientific model is introduced in a discovery process it is instead exactly the only object we plausibly *know* to exist (for example a diagram in a blackboard, or a in-vitro artifact, or a mental imagery). Only in the framework of an oldfashioned strong metaphysical realism we can state that, once a final scientific result has been achieved, together with the description of the related

[7] Cartwright (1983), more classically and simply, speaks of "prepared description" of the system in order to make it amenable to mathematical treatment.

experimental side, everything that does not fit that final structure is a fiction, and so models that helped reach that result itself. Morrison is pretty clear about the excessive habit of labeling fictional scientific models simply because they are superficially seen as "unrealistic": "Although there is a temptation to categorize any type of unrealistic representation as a 'fiction', I have argued that this would be a mistake, primarily because this way of categorizing the use of unrealistic representations tells us very little about the role those representations play in producing knowledge" (Morrison 2009, p. 133).

In the framework of an account of scientific representation in terms of partial structures and partial morphisms Bueno and French (2011, p. 27) admit that they agree in fact that an important role for models in science is to allow scientists to perform the so-called "surrogative" reasoning, but they add the following constraint: "Indeed, we would claim that representing the 'surrogative' nature of this reasoning effectively rides on the back of the relevant partial isomorphisms, since it is through these that we can straightforwardly capture the kinds of idealizations, abstractions, and inconsistencies that we find in scientific models". So to say, we can speak of surrogates, fictions, credible worlds, etc., but it is only through the suitable partial isomorphism we can detect after a success of the model, that we can be assured to be in presence of a scientific representation or model.

Further, Kuorikoski and Lehtinen (2009, p. 121) contend that: "The epistemic problem in modelling arises from the fact that models always include false assumptions, and because of this, even though the derivation within the model is usually deductively valid, we do not know whether our model-based inferences reliably lead to true conclusions". However, the false premises (also due to the presence in models of both substantive and auxiliary assumptions) are not exploited in the cognitive process, because, in various heuristic processes, only the *co-exact* ones are exploited. The notion of co-exact properties, introduced by Manders (2008), is worth to be further studied in fields that go beyond the realm of discovery processes of classical geometry, in which it has been nicely underscored. Mumma (2010, p. 264) illustrates that Euclid's diagrams contribute to proofs only through their co-exact properties. Indeed

> Euclid never infers an exact property from a diagram unless it follows directly from a co-exact property. Exact relations between magnitudes which are not exhibited as a containment are either assumed from the outset or are proved via a chain of inferences in the text. It is not difficult to hypothesize why Euclid would have restricted himself in such a way. Any proof, diagrammatic or otherwise, ought to be reproducible. Generating the symbols which comprise it ought to be straightforward and unproblematic. Yet there seems to be room for doubt whether one has succeeded in constructing a diagram according to its exact specifications perfectly. The compass may have slipped slightly, or the ruler may have taken a tiny nudge. In constraining himself to the co-exact properties of diagrams, Euclid is constraining himself to those properties stable under such perturbations.

Moreover, some false assumptions are considered as such *only if* seen in the light of the still "to be known" target system, and so they appear false only in a *post hoc* analysis, but they are perfectly true in the model itself in its relative autonomy during the smart heuristic cognitive process related to its exploitation. So various aspects of the model are the legitimately true basis for the subsequent exploration of its behavior and performance of the abductions to plausible hypotheses concerning the target system. I agree with Morrison: "I see this not as a logical problem of deriving true conclusions from false premises but rather an epistemic one that deals with the way false representations transmit information about concrete cases" (Morrison 2009, p. 111)[8].

In sum, I think it is misleading to analyze models in science by disregarding the dynamic aspects of the scientific enterprise. Scientific models in a static perspective (for example when inserted in a textbook) certainly appear—but just appear—fictional, because they are immediately compared with the target systems and their complicated experimental apparatuses: in this case also the *ideal* character of models becomes manifest and so the *explanatory* function of them (cf. (Weisberg 2007)). Contrarily, scientific models seen inside the living dynamics of scientific creativity, which is the key topic of epistemology at least since Karl Popper and Thomas Kuhn, appear *explicit* and *reproducible* machineries intentionally built and manipulated to the gnoseological aims of increasing scientific knowledge *not yet available*.

Morrison (2009) is certainly not inclined to see models as fictions because she emphasizes that in science they are specifically related to ("finer graded") ways of understanding and explaining "real systems", far beyond their more collateral predictive capabilities and their virtues in approximating. She indeed further clarifies that the models which is appropriate to label as *abstract* resist—in the so-called process of de-idealization—corrections or relaxing of the unrealistic assumptions (such as in the case of mathematical abstractions or when models furnish the sudden chance for the applicability of equations), because they are "necessary" to arrive to certain results. The fact that in these models "relevant features" are subtracted to focus on a single—and so isolated—set of properties or laws, as stressed by Cartwright (1989), is not their central quality, because what is at stake is their capacity to furnish an overall new depiction of an empirical (and/or theoretical, like in case of mathematics or logic) framework: "[...] We have a description of a physically unrealizable situation that is required to explain a physically realizable one" (p. 130).

Other models, easier to define, which is better to classify as *idealizations*, allow "[...] for the addition of correction factors that bring the model system closer (in representational terms) to the physical system being modelled or described" (Morrison 2009, p. 111). It is for example the case of simple pendulum, where we know how to add corrections to deal with concrete phenomena. Idealizations distort or omit properties, instead abstractions introduce a specific kind of representation "that is not amenable to correction and is necessary for explation/prediction of the target sys-

[8]Further information about the problem of the mapping between models and target systems through *interpretation* are provided by Contessa (2007, p. 65)—interpretation is seen as more fundamental than surrogative-reasoning: "The model can be used as a generator of hypotheses about the system, hypotheses whose truth or falsity needs to be empirically investigated". By using the concept of interpretation (analytically and not hermeneutically defined) the author in my opinion also quickly adumbrates the creative aspects in science, that coincide with the fundamental problem of model-based and manipulative abduction (cf. (Magnani 2009, chapters one and two)).

tem" (p. 112), and which provides information and transfer of knowledge. Morrison's characterization of scientific models as abstract is in tune with my emphasis on models as *constitutive*, beyond the mere role played by models as idealizations, which instead allow corrections and refinements. In this perspective, "abstract" models, either related to prepare and favor mathematization or directly involving mathematical tools, have to be intended as poietic ways of producing new intelligibility of the essential features of the target systems phenomena, and not mere expedients for facilitating calculations. If idealization *resembles* the phenomena to be better understood, abstract models *constitute* the resemblance itself, as I will illustrate in the following subsection.

When Mäki (2009, p. 31) contends that "It may appear that a fantastically unreal feature is added to the model world, but again, what happens is that one thereby removes a real-world feature from the model world, namely the process of adjustment", I have to note that, at least in various creative processes, the model is not necessarily implemented through "removal" or "neutralization" of real-world features, because some features of the target system—that is the supposed to be real world—have simply not been discovered yet, and so, paradoxically, they are the ones still "missing". Consequently it is impossible to imagine that some aspects of the model derive from a removal of features of the real world, that can just be those features that will derive later on exactly thanks to that cognitive process that constructed the model itself to reach that objective. At the same time, and for the same reason, it is difficult to always state that models depict a "surrogate" systems, because the systems we want to subrogate *are largely not yet known.*

3.1 Resemblance and Feyerabend's Counterinduction

Even the concept of resemblance (similarity, isomorphism, homomorphism, etc.) as it is employed in the epistemological framework of missing systems (and related topics, fictions, surrogate systems, credible world, make-believe models, etc.) is in part misleading. "*M* resembles, or corresponds to, the target system *R* in suitable respects and sufficient degrees. This second aspect of representation enables models to serve a useful purpose as representatives: by examining them as surrogate systems one can learn about the systems they represent" (Mäki 2009, p. 32): I contend that resemblance is constitutively partial *also* because it is basically impossible to appropriately resemble things that are not yet known.[9]

It is not always acknowledged in the current literature that isomorphism, homomorphism and similarity with the target systems *are not* necessarily established—so to say—a priori, because the target system has still to be built. Actually—this is an important point—it is just the work of models that of creating, in a poietic way, the "resemblance" to the target system. Some discovered features of the target system

[9]On the puzzling relationships between similarity and representations, in the framework of intentionality, cf. (Giere 2007).

resemble the model not because the model resembled them a priori but only *post hoc*, once discovered thanks to the modeling activity itself, in so far as resemblance has been *instituted* by the model: the new features appear well-defined only in the static analysis of the final developed theory. It is at this stage that resemblance acquires the actual status of resemblance, in the common sense of the word: similarity of two *given* entities/structures. Morrison too contends that "To say that fictional models are important sources of knowledge in virtue of a particular kind of similarity that they bear to concrete cases or systems is to say virtually nothing about how they do that. Instead what is required is a careful analysis of the model itself to uncover the kind of information it yields and the ways in which that information can be used to develop physical hypotheses" (Morrison 2009, p. 123).

In this perspective we paradoxically face the opposite of the received view, in the result of a considerable part of scientific discovery processes it is the newly known target system that resembles to the model, which itself originated that resemblance.[10] Often models are useful to discover new knowledge just because they do not—or scarcely— resemble the target system to be studied, and are instead built to the aim of finding a new general capacity to make "the world intelligible".[11]

In *Against Method* (1975), Feyerabend attributes a great importance to the role of contradiction, against the role of similarity. He establishes a "counterrule" which is the opposite of the neopositivistic one that it is "experience" (or "experimental results") which measures the success of our theories, a rule that constitutes an important part of all theories of corroboration and confirmation. The counterrule "[...] advises us to introduce and elaborate hypotheses which are inconsistent with well-established theories and/or well-established facts. It advises us to proceed counterinductively" (Feyerabend 1975, p. 20). Counterinduction is seen more reasonable than induction, because appropriate to the needs of creative reasoning in science: "[...] we need a dream-world in order to discover the features of the real world we think we inhabit" (p. 29). We know that counterinduction, that is the act of introducing, inventing, and generating new inconsistencies and anomalies, together with new points of view incommensurable with the old ones, is congruous with the aim of inventing "alternatives" (Feyerabend contends that "proliferation of theories is beneficial for science"), and very important in all kinds of creative reasoning. Feyerabend stresses the role of "dreaming", but these dreams are Galileo's dreams, they are not fictions: as I have already pointed out Feyerabend clearly distinguished between scientific dreams (as modeling) and propaganda, that can instead be organized thanks to fictions, inconsistent thought experiments, mistakes, aggressive fallacies, and so on,

---

[10] I endorse many of the considerations by Chakravartty (2010), who stresses the unwelcome division between informational and functional perspective on models and representations in science, which negatively affects the epistemology of scientific modeling.

[11] I am convinced that knowledge about concepts such as resemblance, imaginability, conceivability, plausibility, persuasiveness, credit worthiness (Mäki 2009, pp. 39–40) would take advantage of being studied in the framework of the rigorous and interdisciplinary field of abductive cognition (Magnani 2009), which surprisingly is largely disregarded in the studies of the "friends of fiction", with the exception of Sugden (2000; 2009).

but that do not play any epistemic role in the restricted cognitive process of scientific discovery, I have called "epistemic" warfare (Magnani 2017, chapter four).

Coming back to the problem of models as surrogates, Mäki (2009, p. 35) says:

> The model functions as a surrogate system: it is construed and examined with a desire to learn about the secrets of the real world. One yearns for such learning and sets out to build a model in an attempt to satisfy the desire. Surrogate models are intended, or can be employed to serve, as bridges to the world.

First, I would add some auxiliary notes to the expression "secrets of the real world". I would warn about the preferability of being post-Kantian instead than pre-Kantian by admitting that, through science, we are *constructing* our rational knowledge of the world, which consequently is still an objective world independent of us, but constructed. If we say we build surrogate systems to learn about the secret of nature, a strong realist assumption seems to be presupposed: the models would be surrogates because they are not "reliably reflecting the true reality of the world we are discovering". We rejoin Giere's observation who suspects fictionalists are paradoxically obsessed by "the truth, the whole truth, and nothing but the truth":

> It seems to me that the assimilation of scientific models to works of fiction presupposes an exaggerated conception of nonfiction. On this conception, a genuine work of nonfiction has to provide "the truth, the whole truth, and nothing but the truth". Thus, the realization that scientists are mostly in the business of constructing models that never provide a perfect fit to the world leads to the unwarranted conclusion that scientists are in the business of producing fictional accounts of the world (2009, p. 254).

Scientific theories would reflect this hyper-truth that in turn would reflect true reality (curious! Is not science the realm or self-correcting truths?)[12] In this way it becomes easy to say that everything else in science different from complete established true theories—which would reflect "real world"—is fiction, surrogate, belief, mere credible world, etc.

I would reserve the label of surrogate models to those models employed in some "sciences" that fail in providing satisfactory knowledge about target systems. "There is a long tradition in economics of blaming economists for failing in just this way: giving all their attention to the properties of models and paying none to the relations of the model worlds to the real world" (Mäki 2009, p. 36). Mäki calls the systems described by such models "substitute systems": I will just reserve for them the expression "surrogate systems", because they fake a scientific knowledge that is not satisfactorily achieved, from various perspectives.

I argued above about the epistemological poverty of the concept of model as make-believe: indeed I have already said that make believe processes trivially occur in almost every human intersubjective interplay. Here I can further stress that the idea of credible world is very wide: every cognitive process that aims at providing scientific—but also non scientific—knowledge aims at the same time at providing credible worlds. The problem in science is how to construct the subclass of *epistemologically* credible worlds, that is, *scientific* models, which successfully lead to

---

[12]We should not forget what Morrison reminds us: "Laws are constantly being revised and rejected; consequently, we can never claim that they are true or false" (Morrison 2009, p. 128).

scientific theories. In this spirit Sugden (2009, p. 10) usefully suggests that an epistemologically "good" credible world would have to be provided by models that are able to trigger hypotheses about the "causation of actual events", that is in cases in which "the fictional world of the model is one that *could* be real". Cartwright's classical perspective (Cartwright 2009a) concerning capacities is fruitfully adopted:

> For her, the function of a model is to *demonstrate the reality* of a capacity by isolating it—just as Galileo's experiment demonstrates the constancy of the vertical component of the acceleration of a body acted on by gravity. Notice how Cartwright speaks of *showing that C* has the capacity to produce *E*, and of deriving this conclusion from *accepted principles*. A satisfactory isolation, then, allows a real relationship of cause and effect to be demonstrated in an environment in which this relationship is stable. In more natural conditions, this relationship is only a latent capacity which may be switched on or off by other factors; but the capacity itself is stable across a range of possible circumstances. Thus, the model provides a "theoretical grounding" for a general hypothesis about the world (Sugden 2009, p. 20).

Sugden prudently considers too strong these perspectives on models as tools for *isolating* the "capacities" of causal factors in the real world, and provides other conceptual devices to save various aspects of epistemological—supposed to be weak—"sciences", for example some parts of biology, psychology, or economics, which not ever fulfill the target of revealing capacities. To save these sciences he says that models can simply provide "conceptual explorations", which ultimately contribute to the development of genuinely explanatory theories or credible counterfactual worlds which can trigger inductive (or, better, "abductive") inferences to explain the target systems. I think that it is virtuous to be prudent about strong methodological claims such as the ones advanced by Cartwright, but the epistemological problem remains open: in the cases of models as conceptual explorations are they used to depict credible worlds able to reach satisfactory theorization of target systems, or are they just providing ambitious but unjustified hypotheses, devoid of various good epistemological requisites?

Adopting Cartwright's rigid demarcation criterium clearly and recently restated in "If no capacities then no credible worlds" (Cartwright 2009a), it would seem that no more citizenship is allowed to some post-modern exaggeration in attributing the label "scientific" to various proliferating areas of academic production of knowledge, from (parts of) psychology to (parts of) economics, and so on, areas which do not—or scarcely—accomplish the most common received epistemological standards, for example, the *predictivity* of the phenomena that pertain the explained systems. Are we sure that this demarcation is too rigid or it is time to criticize some excess in the proliferation of models supposed to be "scientific"? It is in this perspective that the epistemological use of the so-called credible worlds appears theoretically suspect, but ideologically clear, if seen in the "military" framework of the academic struggle between disciplines, dominated—at least in my opinion—by a patent proliferation of "scientific" activities that just produce bare "credible" or "surrogate" models, looking aggressively for scientificity, when they actually are, at the best, fragments of bad philosophy.

An example is furnished by the precarious condition of various parts of psychological research. Miller (2010, p. 716) explores three contentions: "[…] that the dominant discourse in modern cognitive, affective, and clinical neuroscience assumes that we know how psychology/biology causation works when we do not; that there are serious intellectual, clinical, and policy costs to pretending we do know; and that crucial scientific and clinical progress will be stymied as long as we frame psychology, biology, and their relationship in currently dominant ways". He further rigorously illustrates the misguided attempts to localize psychological function via neuroimaging and the misunderstandings about the role of genetics in psychopathology, sadly intertwined with untoward constraints on health-care policy and clinical service delivery.

4 Conclusions

In this paper I have contended that scientific models are not fictions. I have illustrated that it is misleading to analyze models in science by disregarding the dynamic aspects of the scientific enterprise: indeed the static perspective leads to an overemphasis of the possible fictional character of models because the creative/factive role of modeling is candidly or intentionally disregarded. I have also argued that other various related epistemological approaches to model-based scientific cognition (in terms of surrogates, credible worlds, missing systems, make-believe) present severe inadequacies, which can be detected taking advantage of a dynamic view of science and of the concept of manipulative abduction. A further way of delineating a more satisfactory analysis of fictionalism and its discontents has been sketched by proposing the concept of "epistemic warfare", which sees scientific enterprise as a complicated struggle for rational knowledge in which it is crucial to distinguish epistemic (for example scientific models) from extra-epistemic (for example fictions, falsities, propaganda) weapons. Finally, I have adopted some thoughts of a classical author, which are of help in dealing with scientific modeling: Feyerabend's useful concept of counterinduction in criticizing the role of resemblance in model-based cognition. In this perspective I have paradoxically reached the opposite of the received view: in the result of a considerable part of scientific discovery processes it is the newly known target system that resembles to the model, which itself originated that resemblance.

Acknowledgements Parts of this article were excerpted from L. Magnani (2012), Scientific models are not fictions. Model-based science as epistemic warfare, in L. Magnani and Li Ping (eds.) (2012), *Philosophy and Cognitive Science. Western and Eastern Studies*, Series "Sapere", Vol. 2, Springer, Heidelberg/Berlin, pp. 1–38, and originally published in chapter two and three of L. Magnani, *The Abductive Structure of Scientific Creativity An Essay on the Ecology of Cognition*. Copyright (2017), Springer, Cham, Switzerland.

References

Bardone E (2011) Seeking chances. From biased rationality to distributed cognition. Springer, Heidelberg/Berlin
Bardone E (2012) Not by luck alone: the importance of chance-seeking and silent knowledge in abductive cognition. In: Magnani L, Li P (eds) Philosophy and cognitive science. Western and eastern studies. Springer, Heidelberg/Berlin, pp 185–203
Bokulich A (2011) How scientific models can explain. Synthese 1:33–45
Boumans MJ (2012) Mathematics as quasi-matter to build models as instruments. In: Weber M, Dieks D, Gonzalez WJ, Hartman S, Stadler F, Stöltzner M (eds) Probabilities. Laws, and structures. Springer, Heidelberg/Berlin, pp 307–318
Bueno O, French S (2011) How theories represent. The British J Gen Philos Sci 62:857–894
Cartwright N (1983) How the laws of physics lie. Oxford University Press, Oxford
Cartwright N (1989) Nature's capacities and their measurement. Oxford University Press, Oxford
Cartwright N (2009) If no capacities then no credible worlds. But can models reveal capacities? Erkenntnis 70:45–58
Cartwright R (2009) Models: parables v fables. Insights 1(8):2–10
Chakravartty A (2010) Informational versus functional theories of scientific representation. Synthese 172:197–213
Contessa G (2007) Scientific representation, interpretation, and surrogative reasoning. Philos Sci 74:48–68
Contessa G (2010) Scientific models and fictional objects. Synthese 172:215–229
Feyerabend P (1975) Against method. Verso, London-New York
Fine A (2009) Fictionalism. In: Suárez M (ed) Fictions in science: philosophical essays on modeling and idealization. Routledge, London, pp 16–19
French S (2010) Keeping quiet on the ontology of models. Synthese 172:231–249
Frigg R (2010) Fiction and scientific representation. In: Frigg R, Hunter MC (eds) Beyond mimesis and nominalism: representation in art and science. pp. Springer, Heidelberg/Berlin, pp 97–138
Frigg R (2010) Fiction in science. In: Woods J (ed) Fictions and models: new essays. Philosophia Verlag, Munich, pp 247–287
Frigg R (2010) Models and fiction. Synthese 172:251–268
Giere RN (1988) Explaining science: s cognitive approach. University of Chicago Press, Chicago
Giere R (2007) An agent-based conception of models and scientific representation. Synthese 172:269–281
Giere R (2009) Why scientific models should not be regarded as works of fiction. In: Suárez M (ed) Fictions in science. Philosophical essays on modeling and idealization. Routledge, London, pp 248–258
Godfrey-Smith P (2006) The strategy of model-based science. Bio Philos 21:725–740
Godfrey-Smith P (2009) Models and fictions in science. Philos Stud 143:101–116
Hintikka J (1998) What is abduction? The fundamental problem of contemporary epistemology. Trans Charles S. Peirce Society 34:503–533
Kant I (1929) Critique of pure reason. MacMillan, London. Translated by N. Kemp Smith, originally published 1787, reprint 1998
Kuorikoski J, Lehtinen A (2009) Incredible worlds, credible results. Erkenntnis 70:119–131
Magnani L (2001) Abduction, reason, and science. Processes of discovery and explanation. Kluwer Academic/Plenum Publishers, New York
Magnani L (2004) Conjectures and manipulations. Computational modeling and the extra-theoretical dimension of scientific discovery. Mind Mach 14:507–537
Magnani L (2004) Model-based and manipulative abduction in science. Found Sci 9:219–247
Magnani L (2007) Abduction and chance discovery in science. Int J Knowl-Based Intell Engi 11:273–279
Magnani L (2009) Abductive cognition. The epistemological and eco-cognitive dimensions of hypothetical reasoning. Springer, Heidelberg/Berlin

Magnani L (2012) Scientific models are not fictions. Model-based science as epistemic warfare. In: Magnani L, Li P (eds) Philosophy and cognitive science. Western and eastern studies. Springer, Heidelberg/Berlin, pp 1–38
Magnani L (2017) The abductive structure of scientific creativity. An essay on the ecology of cognition. Springer, Cham, Switzerland
Mäki U (2009) Missing the world. Models as isolations and credible surrogate systems. Erkenntnis 70:29–43
Manders K (2008) The Euclidean diagram. In: Mancosu P (ed) Philosophy of mathematical practice. Clarendon Press, Oxford/New York, pp 112–183
Miller GA (2010) Mistreating psychology in the decades of brain. Perspect Psychol Sci 5:716–743
Mizrahi M (2011) Idealizations and scientific understanding. Philos Stud 160(2):237–252
Morrison M (2009) Fictions, representations, and reality. In: Suárez M (ed) Fictions in science: philosophical essays on modeling and idealization. Routledge, London, pp 110–135
Mumma J (2010) Proofs, pictures, and Euclid. Synthese 175:255–287
Naylor R (1976) Real experiment and didactic demonstration. Isis 67(3):398–419
Newton I (1999) Philosophiae naturalis principia mathematica. In: Cohen IB, Whitman A (eds) General scholium [1726], 3rd edn. University of California Press, Berkeley, CA (1999)
Peirce CS. Collected papers of Charles Sanders Peirce, vol 1-6, Hartshorne C and Weiss P eds.; vol 7-8, Burks AW ed. Harvard University Press, Cambridge, MA, 1931–1958
Portides DP (2007) The relation between idealization and approximation in scientific model construction. Sci Educ 16:699–724
Steel D (2010) Epistemic values and the argument from inductive risk. Philos Sci 77:14–34
Suárez M (2009) Scientific fictions as rules of inference. In: Suárez M (ed) Fictions in science: philosophical essays on modeling and idealization. Routledge, London, pp 158–178
Suárez M (2010) Fictions, inference, and realism. In: Woods J (ed) Fictions and models: new essays. Philosophia Verlag, Munich, pp 225–245
Sugden R (2000) Credible worlds: The status of theoretical models in economics. J Econ Methodol 7:1–31
Sugden R (2009) Credible worlds, capacities and mechanisms. Erkenntnis 70:3–27
Thagard P (2002) The passionate scientist: emotion in scientific cognition. In: Carruthers P, Stich S, Siegal M (eds) The cognitive basis of science. Cambridge University Press, Cambridge, pp 235–250
Thom R (1988) Esquisse d'une sémiophysique. InterEditions, Paris, Translated by V. Meyer, Semio physics: a sketch, Addison Wesley, Redwood City, CA, 1990
Thomson-Jones M (2010) Missing systems and the face value practice. Synthese 172:283–299
Toon A (2010) The ontology of theoretical modelling: models as make-believe. Synthese 172:301–315
Weisberg M (2007) Three kinds of idealizations. J Philos 104(12):639–659
Woods J, Rosales A (2010) Unifying the fictional. In: Woods J (ed) Fictions and models: new essays. Philosophia Verlag, Munich, pp 345–388
Woods J, Rosales A (2010) Virtuous distortion. abstraction and idealization in model-based science. In: Magnani L, Carnielli W, Pizzi C (eds) Model-based reasoning in science and technology. Abduction, logic, and computational discovery. Springer, Heidelberg/Berlin, pp 3–30
Woods J (ed) (2010) Fictions and models: new essays. Philosophia Verlag, Munich
Woods J (2013) Against fictionalism. In: Magnani L (ed) Model-based reasoning in science and technology. Theoretical and cognitive issues. Springer, Heidelberg/Berlin, pp 9–42
Woods J (2018) Truth in fiction: rethinking its logic. Springer, Cham

On Stubbornness and Cognitive Stability in Rhetoric Systems

Chris Mays

Abstract This phenomenon of stubbornness—refusing to reconsider one's belief or point of view in spite of contrary evidence—is a useful illustration of the way that in any situation, there are multiple ways to view the "facts" of an issue. As this paper argues, a study of *rhetoric* reveals that this phenomenon is a result of the existence of radically different but coexisting logical structures and value systems, which often clash when individuals interact. This point leads to a central claim of this paper: that these divergent structures of argument are actually divergent *rhetoric systems* that *profoundly impact our cognition about the world*. Specifically, a rhetoric system is a set of interconnected rhetorical elements (beliefs, arguments, commonplaces [*loci communes*], meanings, and texts) that cohere into a self-organized system. The borders of this system are stabilized such that, to a person caught up in a particular rhetoric system, (1) his or her cognition pertaining to specific events, texts, and words will seem static, (2) the total holistic understanding of those static meanings will be internally coherent, and (3) the system will be constrained by the other rhetorical, social, biological, and physical systems with which it is in relation. In this sense, a rhetoric system is a "body" that is thoroughly "about" its contexts, and that helps shape cognitive processes. To illustrate these points, the paper examines a particular belief that seems strongly resistant to change: the belief that the "founders" of the United States were devout Christians. As the paper argues, what are often considered stubborn beliefs, rather than exclusive features of an ideology or political party, are actually an integral and complex part of the way all cognition (and knowledge) is created and sustained in a rhetoric system.

C. Mays (✉)
University of Nevada, Reno, Nevada, USA
e-mail: cmays@unr.edu

T. Bertolotti (ed.), *Cognition in 3E: Emergent, Embodied, Extended*,
Studies in Applied Philosophy, Epistemology and Rational Ethics 56,
https://doi.org/10.1007/978-3-030-46339-7_6

1 On Stubbornness and Cognitive Stability in Rhetoric Systems

In the fall of 2008 *New York Times* journalist David Rohde, along with Afghan journalist Tahir Luddin and their driver Asad Mangal, were kidnapped by pro-Taliban forces in Southern Afghanistan. In a harrowing account of his seven-month captivity and subsequent dramatic escape, Rohde describes on several occasions his efforts to bargain with his captors for his and his two fellow captives' release, only to be met with either false promises or outright refusals. One of the striking things about the whole situation is that so many of Rohde's efforts to negotiate with his antagonists resulted in complete failure:

> During our time as hostages, I tried to reason with our captors. I told them we were journalists who had come to hear the Taliban's side of the story. I told them that I had recently married and that Tahir and Asad had nine young children between them. I wept, hoping it would create sympathy, and begged them to release us. All of my efforts proved pointless (2009).

The vast majority of his article contributes to an overall sense of the futility of his attempts at argumentation. While his guards did interact with him, the information he attempted to convey seemed to have no effect.

At one moment in the story, Rohde describes his captors listening to Western media reports, which routinely presented information describing the ordeal of other kidnapped Westerners, many of whom had not even been directly involved in the conflict in the region. As Rohde writes, "the guards dismissed whatever information did not meet their preconceptions." No matter what they heard, from Rohde or from any other source, his kidnappers refused to consider any information or arguments that did not conform to what they already believed.

In a related story: In a study conducted in the fall of 2005 and the spring of 2006 by political scientists Brendan Nyhan and Jason Reifler, participants were presented with mock newspaper articles that repeated one of several common beliefs about current events—for instance, that the regime of Saddam Hussein was harboring Weapons of Mass Destruction (WMDs) somewhere in Iraq. This information was immediately followed (for some participants) by a correction that unequivocally renounced the previous claim. As Nyhan and Reifler discovered, in cases where the original information confirmed the participants' previous beliefs, the subsequent counterevidence *had no effect* on those beliefs; as the authors write, these participants "failed to update their beliefs when presented with corrective information that runs counter to their predispositions. Indeed, in several cases, we find that corrections actually *strengthened* misperceptions among the most strongly committed subjects" (2006, p. 304; emphasis in original).

As the authors point out, there is "extensive literature in psychology that shows humans… tend to evaluate information with a directional bias toward reinforcing their pre-existing views" (pp. 305–07). They later reproduced the effects of this study with the notion of vaccines, showing that individuals who are against vaccination will become more entrenched in their positions when presented with evidence of the efficacy of vaccination.

As all of this shows, information, arguments, and persuasion itself, when they run contrary to one's beliefs, can be quite *in*effective. As it would appear, dismissal of inconvenient evidence is a phenomenon endemic in both Taliban extremists and American citizens alike.

2 Rhetoric, and a Rhetoric System

There is a long tradition of rhetoric scholarship that conceives of it not only as the study of persuasion itself—what Aristotle referred to as "all available means of persuasion" (1984, I.2.1355b20)—but also as an exploration of the presumption that multiple valid perspectives on any given situation exist among differing audiences. We could go back to the Greek sophist Protagoras to find this point, as he held that there are two opposed *logoi* or rational accounts of any situation (Consigney 1996). Rhetorician Chaim Perelman explains this idea in terms of how language itself works. As Perelman writes, "the language upon which argumentation must rely possesses ambiguities that cannot be worked out in advance" (1982, p. 9). Here is a key point: even if potential interlocutors use the same terms to talk about the same topics, for each side those terms may mean different things, since each person will have differing value systems through which terms and topics are filtered. In short, in this way of thinking of a rhetorical situation, two interlocutors will possess different epistemological frameworks for understanding the world, and thus the same situation will be experienced differently by each. While their frame of reference may share certain reference points, even taken-for-granted ideas and beliefs, for each person, may differ in subtle and incompatible ways. Steven Yarbrough, following Donald Davidson on this point, argues that in order for interlocutors to successfully communicate—as he puts it, to "get" each other—they must "come to share" a "similar method of adjusting [their] use of signs" that is "invent[ed] as they interact" (2010, pp. 3–5). In other words, there must be a common ground for communication to occur, and in order to forestall the potential for slippage of meaning, interlocutors must agree on a version of the world where key reference points are more or less the same.[1]

Starting from this presumption that multiple ways of knowing always simultaneously exist in the world helps us understand that in any situation there are multiple versions of the circumstances of an issue, and, importantly, there are radically different logical structures that can coexist in any given situation, which can clash when individuals interact. Within these structures, even "facts" themselves are interpreted

[1]The question of whether it is possible for these reference points to truly match, in other words whether a shared ontology exists or is even possible, has been debated at length, with some scholars on each side of the question, and some arguing for some kind of middle ground wherein each person has access to a particular perspective on reality (Cherwitz and Hikins 1986), some arguing that ontology and epistemology are fundamentally entangled (Barad 2007), and some questioning the value of the very dualist thinking that sponsors such a division (Jordan and Day 2015). In this case, the intricacies of these debates lie outside the scope of this paper.

differently, just as within these different structures there may be different moral values or beliefs deemed paramount, or there may be different meanings derived from the same words or phrases.[2]

This point leads to a central claim of this paper: that these divergent structures of argument are actually divergent *rhetoric systems* that *profoundly impact our cognition about the world*. Specifically, a rhetoric system is a set of interconnected rhetorical elements (beliefs, arguments, commonplaces [*loci communes*], meanings, and texts) that cohere into a self-organized system. Such a system of rhetoric resembles a variety of systems that have been more extensively theorized, such as the organization of an entire society (a social system [e.g. Luhmann 1995]), of a single cell (a biological system [e.g. Maturana and Varela 1980]), or of a variety of physical and chemical systems (e.g. Kauffman 1995). And while these systems can be thought of as always linked to each other, we can also think of each as discrete.

In this sense, thinking of rhetoric as a system involves conceiving of all the elements of language and discourse as an interconnected set of mutually sustaining relations wherein no individual element in the system exists independently, and the stability of each is determined by, and helps determine, the stability of every other element. As such, the totality of these relations contributes to the stability of the whole. At the same time, while we may think of rhetoric in this way as an overarching system, we may also conceive of a "rhetoric system" of individuals—that is, in my rhetoric system, for instance, all of my beliefs, commonplaces, and so on, are interconnected. Importantly, while each person's rhetoric system can be considered in isolation, it is not necessarily the case that each person's rhetoric system must *only* be considered in isolation. Those aforementioned social, biological, and chemical systems are themselves all discrete and yet linked: as one example, a single cell in a brain can be considered an isolated system, but still is networked to the other cells in the brain and so form a coherent system at a larger level of scale. In the same way, a person's rhetoric system is both discrete and linked to the rhetoric systems of members of his or her community, and to the systems of others in more distant communities, and so on, in a mutually sustaining interrelation at different levels of scale. Thus a person can have an "individual" rhetoric system, but that can also be networked together in a rhetoric system of a "community"; in both cases the individual elements of that system will influence and be influenced by each other.

Thus this rhetoric-systems approach fundamentally posits both connection and discreteness. In my rhetoric system, every one of my beliefs, and every commonplace I adhere to, are all connected to and all reinforce each other, and each helps sponsor the cohesion of the whole of my individual belief system. And yet, these elements are not completely isolated from the beliefs and commonplaces of my neighbors and friends (and enemies), the totality of which manifests as a larger system constituent of the rhetoric system of the community. In a rhetoric-systems approach, one may

[2]There has been much debate and discussion about the definition of the term "facts," whether there can be facts can exist apart from human knowers, and what is the nature of so-called "brute" facts (e.g. Fahrbach 2005; Vintiadis and Mekios 2018). For the purposes of this discussion, we will consider facts to be those things that a person or group believes to be indisputably true. To reduce terministic confusion, this paper will refer to facts as simply "beliefs" whenever the contexts allow.

consider an individual system on its own, or consider a larger system encompassing what could be considered multiple individual systems.

A rhetoric system also can be considered as exhibiting qualities of systems extensively theorized within what is known as Wild Systems Theory (WST) (Jordan 2008; Jordan and Day 2015; Jordan and Vinson 2012). In WST, systems such as neurons, brains, and behaviors constitute nested systems of self-sustaining work (Streeck and Jordan 2009). Consequently, rhetoric systems as well are systems nested within other self-sustaining systems, including other bodily and behavioral systems as well as other rhetoric systems with which such a rhetoric system is in relation. Such thinking transcends the traditional body/mind dichotomy and allows one's view of "bodies" and "borders" to be much more diverse; that is, all bodies—including rhetoric systems (that themselves can be considered rhetorical "bodies")—are *embodiments of their contexts* (see also: Jordan 2008; Jordan and Mays 2017). These bodies function as assemblages that are "held together" (i.e., sustain borders) via differing forces: some via the strong forces of nature (e.g., atomic forces for physical systems), some via gravity (e.g., solar systems and objects on planets), some via social forces (e.g., social systems), and (for rhetoric systems) via a rhetorical force that suffuses the entirety of rhetorical action and deliberation (more on this latter force in a moment).

Furthermore, this paper proposes that a rhetoric system functions as an *autocatalytic* system in that the system is able to sustain itself in a relation to its environment. J. Scott Jordan describes the basic functions of such a system in this way:

The work taking place among such a multi-scale nested system is recursively self-sustaining, what Bickhard (2001) refers to as recursively self-maintaining. As a basic example, when one engages in the behavioral work of consuming an apple (i.e. finding it, picking it and eating it) the behavioral work (i.e. finding, picking and eating) produces a product (i.e. the release of chemical energy in the apple) that sustains the systems (e.g. neurons, neural networks and muscles) that made the behavioral work possible in the first place (2008, p. 1983).

Previously, Jordan and Ghin (2006), working within a WST framework, attribute autocatalytic self-sustainment to single cells, since a cell affects sustainment-producing changes to its relationship with the environment (i.e. it attains sustenance) in such a way that the micro-macro synergy between the organism and its environment produces products that sustain the work of this synergy. As well, Jordan notes that Skinner (1976) recognizes behavioral work as self-sustaining, since "behaviors are maintained within an organism's repertoire as a function of the products (i.e. outcomes) they produce, with reinforcing and punishing outcomes being sustained and deselected, respectively" (2008, p. 1983). Jordan and Wesselmann (2014) extend this point to argue that a person's moral behavior is self-sustaining in a social group, and is "about" its contexts on multiple levels of scale. Similarly, Jordan and Heidenreich (2010) maintain that a neuron is self-sustaining by being "about" its contexts on multiple levels of scale.

In much the same way, I argue that a rhetoric system is "about" its contexts, both rhetorical and non-rhetorical. A system being about its contexts, in short, means that there is no separation between system elements and context—that is, system elements are constituted by those very contexts within which they exist and of which they are a part. As Jordan and Vinson put this point, "there is no epistemic gap" between a system and its environment (2012, p. 9). Crucially, this means that elements are determined by and emerge differently according to the contexts in which they exist. In a rhetoric system, the contextual affordances and constraints will constitute the rhetorical elements that emerge, and so one's ideas, beliefs, knowledge, and so on will be constrained and enabled by their contexts, with different contexts giving rise to (potentially) drastically different beliefs and knowledge.

This view of complex systems entails that so-called "physical" systems can constitute constraints of so-called "discursive" ones, a premise that follows in the tradition of applying the tenets of nonlinear dynamic systems both to human society as well as to physical or biological systems (see for example Prigogine and Stengers 1996). In the case of a rhetoric system, the beliefs and meanings that compose the system give rise to lines of reasoning that self-reinforce the cohesion of these beliefs. This, in essence, is the rhetorical force that effects self-sustainment. While there may be change occurring within the system, for example when individual beliefs change, or when values evolve, such change typically is *compensatory* insofar as the changes leave the holistic structure and the boundaries of the system intact. With this kind of change, the internal elements may shift or evolve, but the *borders* of the system—that which defines it as a cohesive worldview—remain essentially stable. Ironically, then, because of this compensatory change, the system is highly resistant to massive destabilization. Such difficult-to-destabilize rhetoric systems become salient on occasions when interlocutors clash over seemingly incompatible reasoning.

Such stability in a rhetoric system leads to precisely the kind of phenomenon on display in the opening examples: that is, stubbornness. This kind of system feature is manifested in several ways. Take, for example, the study on false information about WMDs in Iraq. The dismissal of corrections to misperceptions in this case might easily come from a rationalization that the source of the corrections—the *New York Times*, say—is itself unreliable. With each new challenge to the view, a countermeasure occurs in the system, and another belief is either marshalled or slightly altered to compensate for the potential perturbation of the integrity of the system. Such is precisely the mechanism proposed by Maturana and Varela (1980), who argue that a system "generates and specifies its own organization" in large part by undertaking a "compensation of perturbations" (p. 79).

In other words, external events trigger changes on the inside of the system that allow its continued stability (see also Hayles 1999, pp. 10–11). With regard to a system of rhetoric, events that happen out in the world are thus interpreted solely within a pre-existing network of beliefs; upon hearing some new piece of information I make sense of it completely within the context of what I already know (within the boundaries of my rhetoric system). In this way, nothing on the outside, then, can directly upend the functioning of my rhetoric system inside of its boundaries.[3]

[3]Such a process of reorganization via rationalization also builds on Worsham's idea of discursive reasoning as a "defense mechanism" against "difficult" issues (2013, p. 412); this point also builds on Mays's (2013) argument that this "defensive" process is a primary function of rhetoric itself.

3 Implications of Networked Systems

The premises discussed so far stipulate that the borders of a rhetoric system are stabilized such that, to a person caught up in a particular rhetoric system, (1) his or her cognition pertaining to specific events, texts, and words will seem static, since the system is self-sustaining (2) the total holistic understanding of those static meanings will be internally coherent, and (3) the system will be constrained by the other rhetorical, social, biological, and physical systems with which it is in relation.

This third point, however, needs a bit more unpacking. In short, as previously discussed, a rhetoric system is a "body" that is thoroughly "about" its contexts. Its contexts in this case constitute a variety of systems and elements both discursive and non-discursive, and thus these disparate systems will all act as affordances and constraints on a given rhetoric system. One implication of this is that beliefs ostensibly far-removed from each other are networked. For instance, my beliefs about politics are networked to my beliefs about things as disparate as global warming, where to vacation, the phenomenon of train travel, and even whether to choose paper or plastic bags for items purchased at the grocery store.

Moreover, as mentioned, this networking extends to nondiscursive systems as well. That is, our beliefs are not just influenced by our other beliefs, and the beliefs of our community (though they surely are influenced by both of these), but also by things such as our blood sugar, which affects both our moods and can also directly influence our decision making. To this point, a study by Danziger et al. (2011) found that judges were less likely to grant prison inmates parole when they deliberated right before lunch—a time when presumably their blood sugar levels were the lowest. Additional factors that influence our beliefs include thing like our geographic location, or our specific participation in an economic system, or our material wealth, the weather and climate of where we live, and so on. So, while my arguments about, say, global warming are contingent on the weather patterns I experience, those arguments are also to some extent a product of my geographical residence: if I were a resident of the state of Texas, that is, I might have different beliefs than if I were a resident of the state of Vermont. Such influence extends far beyond linear or obvious relationships, too. It is suffice to say that our rhetoric systems are about the entirety of our environment, in all its manifestations.

One important takeaway of a rhetoric system's quality of radical networking with its environment is that elements like values and even facts can vary and evolve according to perturbations and shifting contexts, creating backwaters of belief that seem apart from the dominant current of contemporary knowledge, or that seem to run directly counter to accepted truths, facts, and values. Moreover, such seemingly faulty (to mainstream thought, at least) knowledge formations themselves exist in rhetoric systems that seem to be cut off from other systems, but in actuality are just as much networked with other systems as any other rhetoric system. As the next case study shows, such seemingly anti-establishment knowledge is just as interconnected, and just as "about" its contexts, as are more mainstream rhetoric systems.

To illustrate these points, the final portion of this paper examines a particular belief that seems strongly resistant to change: the belief that the "founders" of the United States were devout Christians. As the paper argues, what are often considered stubborn beliefs, rather than exclusive features of an ideology or political party, are actually an integral and complex part of the way all cognition (and knowledge) is created and sustained in a rhetoric system.

4 Case Study: The Founders as Fundamentalists Versus the Founders as Skeptics

There is a vocal group of Americans who believe that the authors of key foundational texts of US government, such as the Constitution and the Declaration of Independence, were fundamentalist Christians, who intended for the United States to be a distinctly Christian nation. To those who disagree with them, members of these groups seem to ignore strong counter-evidence that the founders were neither specifically evangelical, nor even clearly religious. One prominent figure in this group is historian and evangelist David Barton, whose own online bio proclaims that "*Time Magazine* called him 'a hero to millions,'" including "some powerful politicians." Barton's views include premises that would seem quite radical to a person adhering to the mainstream view of American history. Various profiles attribute to him such ideas as: "the Founding Fathers were deeply religious men who built America on Christian ideas" (Hagerty 2012), and that "Jesus did not like the minimum wage" (Coaster 2011). Importantly, Barton has been repeatedly criticized for "factual errors" in his work, and his *New York Times* bestseller on Thomas Jefferson was pulled from the shelves by the publisher because of such errors (Hagerty 2012).

On the other side of things, there is the version of history that features the belief that the founders were *not* devout Protestants nor even particularly devout Christians, and were either a-religious, deists, or in general lacked the fervor of devout religious adherents. This view claims that the founders were largely a-religious, and on the whole did *not* intend for religion to be a primary component of government. University of Chicago law professor Geoffrey Stone (2008) outlines one strong example of this view, writing that the "[f]ounding generation viewed religion, and particularly religion's relation to government, through an Enlightenment lens that was deeply skeptical of orthodox Christianity" (p. 8). As *New York Times* reporter Russell Shorto (2010) sums up, while "the founders were rooted in Christianity," they "at the same time… were steeped in an Enlightenment rationalism that was, if not opposed to religion, determined to establish separate spheres for faith and reason."

This latter set of beliefs circulate in a particular rhetoric system from which emerges what historian William Hogeland (2009) calls "'public history'—the history we encounter in museums and tourist attractions, in newspaper columns and election campaigns, in public broadcasting and popular biographies." Such history, Hogeland writes, is often dominated by a "mainstream" historical narrative (pp. xii–xiii). In this case, the mainstream narrative (and the claims that emerge from this narrative) constitute the rhetoric system currently dominant in American political discourse, and, this narrative is significantly opposed to any counterclaims about the founders' evangelical—or even religious—nature.

Barton's texts, which are clearly very influential in some circles, reveal a rhetoric system that to outsiders is often difficult to comprehend, let alone analyze, since it is based on ideas deemed by many as simply not true. In this sense it is easy to dismiss Barton's views as a product of fringe discourse, influential only within insular groups determined not to listen to contradictory mainstream evidence. However, although Barton's texts may seem ignored by a large segment of the public, since rhetoric systems are "about" their contexts, *all* rhetoric systems are about these other systems in subtle but often important and influential ways, and thus Barton's ideas often end up influencing public discussions and public policy significantly.

For example, writing on the Encyclopedia Britannica blog (the slogan of which reads, not incidentally, "facts matter"), Joseph Ellis (2007) explains that "in recent decades Christian advocacy groups, prompted by motives that have been questioned by some, have felt a powerful urge to enlist the Founding Fathers." He concedes that while "recovering the spiritual convictions of the Founders, in all their messy integrity, is not an easy task," what he calls the "dominant pattern" is their "diversity" of belief.

Ellis's points here seem to be a simple acknowledgement of the debate on the issue. But considering this statement in light of the key tenets of Barton's arguments reveals that the ideas portrayed in the Britannica text *have changed to accommodate the views of people who support Barton.* Specifically, Ellis's acknowledgment that the founders' views might have been more complex than is commonly acknowledged in the dominant view of history can be traced to the mainstream salience of views like Barton's. In a further tacit endorsement of this diversity, Britannica sanctions a separate blog (Novak and Novak 2007) that serves as a counterpoint to Ellis: on this blog the authors argue that "in their brief asides on the subject," "virtually all" of the founders agreed with "most Christians even today," who "regard immortal life as a communion with God, with their friends, and all those historical greats whom they admire—an everlasting conversation."

Such influence is evidence that even mainstream rhetoric systems that would seem to ignore outlier views are nevertheless "about" them as well. To illustrate further: During a 2012 event for the US Republican primary election in the state of Iowa called the "Thanksgiving Family Forum," two Presidential candidates (Michele Bachmann and Rick Perry) explicitly endorsed the premise that, as Bachmann put it, "American exceptionalism is grounded on the Judeo-Christian ethic, which is really based upon the 10 Commandments" (Saletan 2011). As Bachmann stated bluntly in an interview: "The Judeo-Christian heritage isn't a belief. It's a fact of our nation's history. It's a fact" (Brody 2010). Historical arguments aside, these are premises pulled directly from Barton's work—and Bachmann freely admits that this is the case. The Congresswoman is an outspoken admirer of Barton's, even going so far as to invite him to be a guest speaker at her proposed weekly "class[es] on the Constitution," which she planned to make available to all Congresspersons in 2010

(Brody 2010). In these very public statements and actions, we can see the beginning of an evolution of the broader public discussion—and the mainstream rhetoric system—in that the broader public, rather than simply ignoring or dismissing these claims, is increasingly being asked to recognize, and to contend with, the existence and popularity of Barton's ideas.

Moreover, zooming in on specific elements of these ostensibly far-flung rhetoric systems reveals even more influence, in this case, in the form of circulating elements—somewhat like building blocks of the system—that travel from one system to the next. Specifically, phrases found in the Britannica entry such as "immortal life," and "communion with God" constitute specific elements of Barton's arguments; in this sense the Britannica article is overtly gesturing to points of view allied with Barton's. Even if the writing on this second Britannica blog does not explicitly reference nor completely parallel Barton's larger arguments, it certainly reproduces a significant part of Barton's rhetoric system structure simply by acknowledging his point of view. Importantly, this interrelation does not mean that the Britannica entry is directly reproducing Barton's work in total. It doesn't have to. Rather, in the spread of these key elements to various discursive sites such as the Presidential debates, classes offered to Congresspersons, and blogs and counter-blogs officially sanctioned by Encyclopedia Britannica, Barton's views, which may at first glance seem confined to only the most esoteric of rhetoric systems, can be seen to clearly interact in significant ways with a wide variety of rhetoric systems.

As one more example of an interaction that a rhetoric-systems approach can highlight: In a 2010 debate over the founders' Christianity that played out in a debate at the Texas Board of Education, compensatory reactions can be found at work in the responses of the board members. As a *New York Times* article on the debate illustrates, the views of members who advocate for the importance of religious fundamentalism appear stable, despite the objections and rhetorical efforts of those opposed to them. However, these board members' responses to criticism from the "secularists" demonstrate that these individuals—ostensibly stubborn and set in their views—have actually changed positions slightly, if only to further rally around their cause. As one of them put it: "You can't appreciate the founding of our country without realizing that the founders understood [these religious truths]. For our kids to not know our history, that could kill a society. That's why to me this [debate] is a huge thing" (Shorto 2010). This quotation may sound like increased stubbornness—and in a sense, it is. But it is also a change in the dynamic of the debate, and illustrates how rhetoric systems are thoroughly about those other systems in their environment, often in such a way that illuminates the compensatory adaptions of complex systems. That is, as the quotation above shows, those who hold a radical view of the religiosity of the founders have adopted a position that remains contentious, but that also recognizes that criticism of their views has become widespread. Just as the belief that the founders were Christian has become widely circulated in the mainstream rhetoric system, so has criticism of that view become more widely circulated in this other rhetoric system. Although the board members quoted in this *Times* article argue that failure to teach kids their version of history could "kill a society" (which sounds like an indication of increased stubbornness and high stability), these same individuals' tone of urgency in response to a perceived threat to their worldview is itself a manifestation of an evolution of their views.

5 Rhetoric Systems as Constraints, and as Constrained by the World

A central implication of the view of rhetoric systems articulated in this paper is that these systems are fundamentally about all other systems in their environment, and vice versa. This means that these systems serve as constraints on other systems, and themselves are constrained by those same systems, and both physical and discursive systems are implicated in this interactive relationship.

In short, rhetoric systems have a significant impact on our cognition. Both by compensatory reaction and by other adaptations we incorporate, adjust to, and defend against elements and changes in other systems, and in so doing our way of thinking, what we value, and what we consider bedrock facts or beliefs of our worldview are all impacted. While this impact may at times be difficult to detect, it is important to recognize the way our cognition and our worldview, even while seeming coherent and stable, are subtly but continually moving as contexts and surroundings evolve.

Reference

Aristotle, Roberts R, Bywater I (1984) The rhetoric and the poetics of aristotle. Random House

Barad K (2007) Meeting the universe halfway: quantum physics and the entanglement of matter and meaning. Duke University Press

Bickhard MH (2001) The emergence of contentful experience. In Kitamura T (eds). What should be computed to understand and model brain function? World Scientific, (pp 217–237)

Brody D (2010) Gingrich, Bachmann speak to brody file about new GOP congress. Brody File. Christian Broadcasting Network

Cherwitz RA, Hikins, JW (1986) Communication and knowledge: an investigation in rhetorical epistemology. University of South Carolina Press

Coaster L (2011) Rev. David Barton was Michele Bachmann's history teacher. Dailey Kos, Kos Media

Consigny S (1996) Edward schiappa's reading of the sophists. Rhetoric Rev 14(2):253–269

Danziger S, Levav J, Avnaim-Pesso L (2011) Extraneous factors in judicial decisions. Proc Natl Acad Sci USA 108(17):6689–6892

Ellis J (2007) David Barton Bio. (n.d.) Wall Builders. The U.S. Founding Fathers: Their religious beliefs. Encyclopedia Britannica Blog. Encyclopedia Britannica, Inc

Fahrbach L (2005) Understanding brute facts. Synthese 145(3):449–466

Hagerty BB (2012) The most influential evangelist you've never heard of. NPR

Hayles NK (1999) How we became posthuman. University of Chicago Press

Hogeland W (2009) Inventing american history. MIT Press

Jordan JS (2008) Wild-agency: nested intentionalities in neuroscience and archeology. Philos Trans Roy Soc London B: Biol Sci 363(1499):1981–1991

Jordan JS, Day (2015) Wild Systems Theory as a 21st century Coherence framework for cognitive science. In: Metzinger T, Windt JM (eds), Open MIND: 21(T). MIND Group

Jordan JS, Ghin M (2006) (Proto-) consciousness as a contextually-emergent property of self-sustaining systems. Mind Matter (4):45–68

Jordan JS, Heidenreich B (2010) The intentional nature of self-sustaining systems. Mind Matter 8(1):45–62

Jordan JS, May C (2017) Wild meaning: the intercorporeal nature of objects, bodies, and words. In: Meyer C, Streeck J, Jordan JS (eds). Intercorporeality: emerging socialities in interaction, Oxford University Press, (pp 361–378)

Jordan JS, Vinson D (2012) After nature: on bodies, consciousness, and causality. J Conscious Stud 19(5/6):229–250

Jordan JS, Wesselmann ED (2014) The contextually grounded nature of prosocial behavior: a multiscale, embodied approach to morality. In: Schroeder DA, Graziano WG (eds). The oxford handbook Of prosocial behavior. Oxford University Press

Kauffman S (1995) At home in the universe: the search for the laws of self-organization and complexity. Oxford University Press

Luhmann N (1995) Social Systems. Trans. Bednarz Jr J with Baecker D. Stanford University Press

Maturana HR, Varela FJ (1980) Autopoiesis and cognition: the realization of the living. Reidel

Mays C (2013) Who's driving this thing, anyway?: emotion and language in rhetoric and neuroscience. JAC 33(1–2):473–496

Novak M, Novak J (2007) The god of liberty and the U.S. founding fathers. Encyclopedia Britannica Blog. Encyclopedia Britannica, Inc

Nyhan B, Reifler J (2006) When corrections fail: the presistence of political misperceptions. Paper presented at the annual meeting of the American Political Science Association, 303–330

Perelman C (1982) The realm of rhetoric. University of Notre Dame Press

Prigogine I, Stengers I (1996) The end of certainty: time, chaos, and the new laws of nature. Simon and Schuster

Rohde D (2009) Held by the Taliban: a times reporter's account. A five-part series by David Rohde. Nytimes.com, The New York Times Company

Saletan W (2011) Rule of lord: the republican plan to nullify the courts and establish christian theocracy. Slate, Washington Post Company

Shorto R (2010) How christian were the founders? Nytimes.com, The New York Times Company

Skinner BF (1976) About behaviorism. Vintage Books. Sprague RK ed (1972). *The older sophists*, University of South Carolina Press

Stone GR (2008) the world of the framers: a christian nation? Univ Calif Law Rev 56(1):1–26

Streeck J, Jordan JS (2009) Communication as a dynamical self-sustaining system: the importance of time-scales and nested contexts. Commun Theor 19(4):445–464

Vintiadis E, Mekios C (eds) (2018) Brute facts, Oxford University Press

Worsham L (2013) From the editor: Thinking with Cats (more, to follow). JAC J Rhetoric Cult Politics 30(3–4):407–433

Yarbrough S (2010) On "getting it": resistance, temporality, and the "ethical Shifting" of discursive interaction. Rhetoric Soc Q 40(1):1–22

From Emotions to Artifacts: Four Modes of Fulfilling Life-Relevant Tasks

Marco Viola

> *Fear is a superpower. Fear can make you faster, and cleverer, and stronger*
> Clara, Doctor Who, S08, E04
> *Fear. Somedays it feels like there is no escaping it. But there is one place where fear has no place: home. SimpliSafe stops fear at the front door.*
> 'A world full of fear', commercial for the home security systems Simply Safe.

Abstract Following the taxonomy of four modes of cognition sketched by Roberto Casati (a syncretism of dual system theories of reasoning and distributed cognition), I describe four modes of addressing evolutionary salient tasks, i.e. those tasks like foraging and mating on which the survival of the species hinges upon. According to an influential tradition, (1) emotions are toolkits designed by evolution to efficiently address biologically relevant challenges. However, Homo sapiens address some of these challenges by (2) rearranging the workings of emotions by means of higher-cognitive faculties. Moreover, still other challenges are dealt with by (3) scaffolding or even totally (4) offloading the relevant tasks to external resources, such as artifacts.

Once upon a time, Earth was only inhabited by unicellular organisms. In order to survive, they had to react to the environment to avoid harmful chemicals in favor of nutrients of some sort. Only the organisms that accomplished this trivial task had some chance of survival and evolution into more complex organisms. Ultimately, after a long and non-linear evolutionary journey, currently living species emerged, including human beings. As organisms grew in complexity, so did the challenges they must face. Besides avoiding bad chemicals and looking for good ones, the survival of both the self and the species requires to fulfil further tasks, such as avoiding other noxious organisms, which may be big (such as predators) or small (such as

M. Viola (✉)
Department of Philosophy and Education, University of Turin, Via Sant'Ottavio 20, Turin, Italy
e-mail: marco.viola@unito.it

T. Bertolotti (ed.), *Cognition in 3E: Emergent, Embodied, Extended*,
Studies in Applied Philosophy, Epistemology and Rational Ethics 56,
https://doi.org/10.1007/978-3-030-46339-7_7

pathogens); finding one (or more) suitable partner(s); protecting and rearing the offspring; or physically overcoming a rival or a prey (LeDoux 2012).

According to an influential tradition, the mental states we call emotions are the product of cognitive toolkits designed by evolution to address these biologically relevant challenges efficiently. For instance, fear facilitates the flight from predators, disgust prevents us from ingesting pathogens, lust promotes mating, love and jealousy make relationships more stable, tenderness and care prompt nurturing, while anger facilitates in times of struggle—to name but a few.

However, evolution is a slow designer: the toolkits that we have nowadays were selected because of their usefulness in facing environmental challenges that differ significantly from those we encounter nowadays. Were we to rely on our old-fashioned disgust reactions, we might end up turning down broccoli (perhaps in favor of some sweet junk food); were we trying to address all our enmities with a fight, we would probably go to jail. Fortunately, we are flexible enough to learn some strategies that complement our emotional toolkits: with a bit of education we learn to avoid eating too much sugar, while with a bit of cunning we learn to defer our petty revenges from the here and now in favor of something more cerebral.

But the story does not end here. Modern biologists (or at least some of them) no longer see evolutionary forces as proceeding linearly from the environment to the organisms, with no feedbacks in the opposite direction. Rather, they acknowledge that humans (and other animals) engineer their own niche to 'cheat' out of undesired evolutionary pressures (Laland et al. 2000). For example, humans learned to use tools to go beyond the limitations of their bare bodies, and the invention of weapons and traps enabled our ancestors to hunt larger preys. Moreover, some tools also allow us to go beyond our bare minds, scaffolding our cognitive capacities: for instance, we offload mnemonically burdensome information to notebooks (Clark and Chalmers 1998; Sterelny 2010). I argue that some of these environmental scaffolds concern the same kind of life-relevant tasks for which emotional toolkits originally evolved, and work either by complementing or by completely subsiding them. In a sense, they are *prosthetics or mirror images of human emotional systems*.

In this essay, I set out to take this analogy seriously, by suggesting that the functions typically exploited by our emotions have been offloaded to artifacts as well as to other environmental scaffolds that vicariate and extend them. To do so, in the next section I introduce Roberto Casati's recent discussion in which he distinguishes four modes of performing cognitive tasks. His account builds on the dual system theory in psychology of reasoning (made extremely popular by Kahneman's 2011 book *Thinking Fast and Slow*), and connects it to the literature on cognitive artifacts. Then, in the following section, I discuss four modes of addressing life-relevant tasks. This parallel sheds light on some implications concerning how we (should) conceive of our emotions and our (extended/scaffolded) mind, which I finally address in the concluding section.

1 Casati on Four Modes of Cognition

A couple of decades ago, some researchers within psychology of reasoning proposed a distinction between two (kinds of) cognitive systems underlying thinking (Stanovich and West 2000). The Nobel Prize for economics Daniel Kahneman, who greatly contributed to the distinction's popularity, describes the two systems as such:

> System 1 operates automatically and quickly, with little or no effort and no sense of voluntary control. System 2 allocates attention to the effortful mental activities that demand it, including complex computations. The operations of System 2 are often associated with the subjective experience of agency, choice, and concentration (Kahneman 2011, p. 20–21).

Many mental tasks may be undertaken by either system: to put it bluntly, switching from System 1 to System 2 usually implies a tradeoff between speed and accuracy. However, Kahneman urges us against reifying system-talk, by stressing that these systems should be seen as "fictitious characters [...] not systems in the standard sense of entities with interacting aspects or parts. And there is no one part of the brain that either of the systems would call home" (*ibid.*, p. 32).

To discourage reification, Roberto Casati (2017) recently replaced the theoretically controversial term "systems" with the relatively noncommittal "modes" (shortened in M1 and M2). His aim is to expand this taxonomy by including some insights from the debates on distributed cognition, while introducing two further modes of exploiting cognitive tasks. In what Casati calls Mode 4 (M4), a cognitive task (or at least its core computational processes) is completely offloaded to some cognitive artifact,[1] as in the famous case of the imaginary Alzheimer patient Otto, who recalls the address of the Modern Museum of Art by writing it down on his trustworthy notebook (Clark and Chalmers 1998). M4 is an extreme case, which Casati introduces mainly to better characterize Mode 3 (M3). Tasks undertaken in M3 are only partially offloaded: in such cases, a cognitive artifact is doing *part of the computational job*, somewhat modifying the computational architecture of the (remainder of the) task which is undertaken by a human agent (Norman 1991). While Casati does not make the connection explicit, M3 would probably correspond in most cases to what Sterenly (2010) calls *scaffolded mental processes* (of which Clark and Chalmers' *extended* mental processes are an extreme case).

Rather than proposing a complex definition, Casati clarifies his construal of the four modes through an evocative example: i.e., navigation, that is the process of orienting oneself in space to reach a particular destination.

For most people, driving home from the office is a rather effortless task: they know where to make a turn and when to stop. They might decide to put up a Ted

[1]Traditionally, the term 'artifact' is restricted to those tools that have been intentionally designed for a specific purpose (e.g. a knife is designed for cutting). Non-human-made items that get repurposed for some function they were not originally designed for (e.g. a wooden stick used for drawing on the sand) are sometimes called 'naturefacts'. In this paper, I overlook this and other similar distinctions, and I use the term 'artifacts' more loosely, to refer to every tool or element in the environment that is employed for some purpose.

Talk while they drive, and manage not to get lost while actually listening to it. They are navigating in M1.

By contrast, Casati recalls the first time he drove in England, a left-hand traffic country (unlike his native country). In that situation he felt like a newbie: he had to plan every action carefully, actively (and slowly!) making decisions each time he has to watch or turn in one or the other direction. No Ted Talk—as interesting as it may be–could be attended in such conditions (unless one is willing to risk having an accident). This is an extreme example of how navigating in M2 looks like. But M2 is also at work every time you have to find a place you do not know.

Now, imagine you are on a journey to some unknown destination with a GPS. You do not need to pay too much attention to the road: while driving is up to you, you totally offload any direction-related choice to that powerful device (in Casati's terms, you are only performing some auxiliary computational tasks yourself, whereas the core computations are performed by the GPS). This is navigating in M4.

Finally, if your GPS is broken, or you simply are the kind of person who prefers driving the old-fashioned way, you might be navigating by following a roadmap. In doing so, while you are still making use of a cognitive artifact that does part of the job (namely, storing a trustworthy spatial representation), you still have to do part of the job yourself (namely, performing the core computations pertaining to navigation). However, the part of the computational job you are doing now is *a different part*: rather than building and consulting a mental map, you must locate yourself onto the external one. This is what navigating in M3 might look like.

In sketching this taxonomy, Casati takes the stance of a designer rather than that of a metaphysician: his interest is not that of establishing the external boundaries and the ontological status of mental states (as is Clark and Chalmers' goal, at least in their 1998 paper). Rather, he aims at stressing that many cognitive tasks allow for multiple solutions, while switching the focus on the internal architecture of such tasks and highlighting possible ways in which such architecture could be redesigned.

In the same spirit, in the following section I discuss a particular set of tasks whose prototypical realization in M1 gives rise to what we call *emotional states*, and I examine the alternative strategies for facing them.

2 From Hardwired to Artificial Emotion Systems

2.1 M1: Affect Programs

More than a century ago William James posed the question: "What is an emotion?" in his renown 1884 article. Yet, there is no consensus in sight with respect to the answer (see Scarantino 2016). However, one thing on which many scholars agree is that emotions are tightly related to survival and reproduction. A very influential account, inspired by the tradition of evolutionary psychology, conceives of emotions as biologically hardwired functional states, that is pre-wired computations designed by evolution to maximize fitness at the species level—i.e., to ensure that its members survive and reproduce, thus propagating their genes (notably Tooby and Cosmides 2008. See also Ekman 1992; Lench et al. 2015; Panksepp 1998; Adolphs and Anderson 2018). Two often cited examples within this account are *fear* and *disgust*. By regulating physiological parameters in a way that mobilizes energy, fear enables preys to effectively flee from predators (recall the initial quote from Doctor Who: Fear is a superpower!). Fear also does many other useful things which arguably increase our chances of survival, e.g. it widens our eyelids to enlarge the visual field (Susskind et al. 2008). Alternatively, take disgust, another frequently cited example of such systems. Thanks to its purported function of protecting us from possibly poisonous substances, disgust has been compared to a "guardian of the temple of the body" (Rozin et al. 2008).

Computational packages such as the ones described above are sometimes called *affect programs* (or basic/primary emotions). According to Griffiths (1997), they are the kind of mental states that we prototypically have in mind when we talk about emotions in everyday terms. Authors who defend their existence often disagree about their exact number (usually ranging from 5 to 9), and also about their individuation criteria.[2] Nevertheless, fear, disgust and anger appear in most lists. To avoid any preposterous theoretical commitment, Adolphs and Anderson (2018) maintain that, although such discrete emotion categories exist, we cannot still list them exhaustively. In what follows, I assume their prudential "there are some, but I do not which ones" stance.[3]

These systems have also been compared to Fodorean modules (Griffiths 1997; Faucher and Tappolet 2006), to which they resemble in many respects. For instance, like modules, affect programs are purported to have dedicated neural bases (e.g. Panksepp 1998; Vytal and Hamann 2010; but see Lindquist et al. 2012); and to be set off automatically and inescapably given the proper triggers (Ohman and Mineka 2001), in order to produce a fast and frugal response. Indeed, these two aspects are closely interrelated: in the famous dual route model pioneered by LeDoux (1998), the swiftness of responses, as well as their coarseness, are due to the fact that affect

[2]For instance, Ekman (1992) notoriously builds his list of emotions based on patterns of facial expression whereas Panksepp (1998) builds his own on behavioral and neural homologies. See Ortony and Turner (1990) for an influential criticism.

[3]Of course, some scholars *deny* the very existence of (something like) affect programs. For instance, while acknowledging that emotions are tightly related to survival and reproduction through the regulation of one's internal *milieu*, Lisa Barrett (2017) rejects the existence of hardwired emotion categories, claiming instead that specific emotions are shaped by conceptual knowledge in a top-down fashion, and that there is no one-to-one link between a specific emotion and behavior. LeDoux (2012) allows for the existence of such systems, but proposes to call them *survival circuits*, suggesting that the word emotion, as well as specific emotion words, should only be employed to characterize the phenomenal feeling which sometimes accompany such behaviors. As my focus here is not the phenomenal aspect of emotions but only their functional roles, everything I say in this section can be made consistent with LeDoux's account (although not with Barrett's) by replacing the expression "emotional system" with "survival circuit".

programs systems (like M1) rely on a sub-cortical route, which bypasses the more sophisticated (but slower) information processing of the higher, cortical route.[4]

A point worth stressing here is that such systems are thought to be synchronically impenetrable (i.e., they disregard information coming from other cognitive systems) but not diachronically impenetrable (i.e., their functioning can be influenced by experience). Indeed, emotional reactions are sometimes in contrast with our conscious assessments: people with phobias will react fearfully to some kinds of stimuli even if they do not think they pose an actual threat.[5] People who cannot stand broccoli will hardly eat them, no matter what the doctor recommended.

This immunity from conscious control is a property of emotions that philosophers dubbed *recalcitrance*. After all, evolution is not an omniscient designer, and in the trade-off between false-positives (e.g., hallucinating a predator and start running uselessly) and false-negatives (e.g., failing to spot the predator and thus get killed for *not* running), it shows a strong preference for the former—for obviously good reasons.

However, recalcitrance does not imply that the appraisal mechanisms activating the affect programs cannot become tuned to new triggers over time: in the long run, new triggers can be learnt. Perhaps, even new ways to react can be added to the motor repertoire of the emotional system. Importantly, this learning process does not seem to be working as a *tabula rasa*: rather, it is allegedly biased toward biologically salient stimuli. In their review of several experiments on human and animal models, Öhman and Mineka (2001) showed that fear learning seems to associate with more ease (and less reversibly) to biologically salient stimuli such as snakes and spiders than to biologically irrelevant stimuli such as flowers and mushrooms. Similarly, Rozin and colleagues (2008) claimed that, while the triggers of disgust are mostly acquired, some stimuli such as body waste products are most easily learnt than others.

Through this admittedly non-exhaustive discussion, I have delineated some emotional phenomena and mechanisms which should be familiar to most, and which qualify as good examples of M1 solutions to recurrent life-challenges. Clearly, complex animals such as humans have complex problems, and these in turn require complex solutions: yet, performing evolutionary relevant tasks like mating in contemporary human societies is almost never easy and automatic enough to require (or even allow for) a M1-style solution. Take fear, the M1-solution for keeping us safe: an experiment by Olson and colleagues (2005) suggests that we are relatively more

---

[4]While not uncontested (see Pessoa and Adolphs 2010), the existence of such a low-cortical route would explain the phenomenon of *affective blindsight*, where people suffering from cortical blindness (i.e., who are incapable of processing visual information at a cortical level) are nonetheless capable of processing some emotionally salient information, albeit at a coarser-grained level (Celeghin et al. 2015).

[5]Moreover, some emotional reactions may act as censors of desires: were he insisting in drinking so many White Russians, the Dude from Great Lebowski might grow sick of his beloved cocktail, thus becoming unable to enjoy it again without vomiting. Such cases are often taken as instances of *emotional ambivalence*. For the sake of simplicity, I do not deal with it here.

afraid of people that we perceive as out-groups on an ethnic base.[6] However, as useful as it might have been for the survival of our tribesmen ancestors, our hardwired diffidence toward other ethnic groups is not currently doing us a good service (in fact, it might increase the level of danger of our societies, by fostering racial tensions).

2.2 M2: Complex Emotions

As I mentioned above, complex organisms have complex lives. In these complex lives, they must deal with environments and challenges that mutate much more rapidly than the evolutionary toolbox provided by M1. Were we only relying on our brute emotional instincts, anger will often lead us into fights, fear into fleeing from whatever *prima facie* looks like a threat, lust into sexual intercourses that we might regret later on, and so forth. Fortunately, unlike the famous neurological patient Phineas Gage,[7] we are not doomed to obey to all of our M1 commands. At least in some cases, people might resist the urge to get away from someone because of the color of his or her skin, the urge to sleep with someone they met on Tinder on our first date, or the urge to punch an annoying guy in the face because he mocked them. I am not saying that we may be able to totally shut off our emotional reactions: as Damasio (1994) convincingly argued, this is not how the human mind works. Rather, according to his *somatic maker hypothesis*, we mediate our instinctive feelings by distancing ourselves from the here and there, and rather employ these gut impressions to navigate the desirability of several scenarios which we foresee as potential consequences of our (possible) courses of actions. To put it bluntly, we do not suppress anger by burying our need for revenge in some Kantian moral principle, but rather we invoke other emotional thoughts (e.g., empathy, "perhaps the guy had a bad day"), or we even defer our revenge to another day (following the motto "revenge is a dish best served cold").

Therefore, the kind of framework I am defending with respect to M2 does not imply that M1 emotional systems would get *suppressed.* Instead, they would be *decoupled* from the specific situation and *redeployed* in favor of more complex emotional reactions, mediated by other cognitive and/or emotional processes. As such, the account I propose is roughly consistent with the ones positing complex emotions (e.g. Izard 2011; Griffiths 1997; Panksepp 1998; see also J. Clark 2010).

---

[6]However, this fear-bias was diminished for those who dated people of different ethnicities. To restate, appraisal are plastic.

[7]For those who never heard the story: Phineas Gage lived in the Nineteenth century and became famous due to a unique accident recounted in Damasio's best-selling book *Descartes' Error* (Damasio 1994). While he was working as a railway builder, an explosion pushed an iron bar into the left part of his skull, injuring his left eye and a great part of the left frontal cortex. Oddly, he survived—and even walked to the doctor by himself despite the iron bar in his head! However, following the accident, Gage was never the same again: he became incapable to suppress or negotiate with his raw instincts, resulting in a compromised social life.

M2 comes particularly in handy when it comes to disgust. Our innate taste pushes us to privilege food with high caloric intake such as meat or sweets over food with lower caloric intake such as vegetables. A bliss in prehistoric age, this feature turns into a curse in current Western society, where the large availability of junk food may easily recruit our innate disposition, leading to widespread problems such as obesity. In order to countenance the temptation to eat more than we need, we need to switch toward M2 thinking: for instance, we may imagine some negative consequences associated with weight gain, and the distress that follows might discourage us from eating more.

The fact that so many people suffer from obesity despite the availability of these M2 strategies reminds us that among the defining features of M2 "is that its operations are effortful, and one of its main characteristics is laziness, a reluctance to invest more effort than is strictly necessary" (Kahneman 2011, p. 31). In order to ease this burden and to mitigate the laziness of M2, we have invented several solutions that manipulate the environment to 'hijack' M1, allowing us to deal with life-relevant tasks more efficiently, or more parsimoniously (M3). At times, we come up with ways for not even having to deal with tasks altogether, letting something else do the job (M4). I describe such cases in the following sections.

2.3 M3: Scaffolded Actions

As I repeatedly stress, we live in a complex world. This degree of complexity evolves too fast for our biological evolution alone to catch up. The variety of threats and poisons we should avoid in order to survive constantly changes, as new ones emerge, and old ones become obsolete. How are we supposed to keep up with them? On the one hand, we cannot wait for our biological equipment to evolve, nor we want to experience every possible annoyance first-hand (M1-learning) only to update the triggers of fear and disgust. On the other hand, we can neither devote all our mental energy to figure out how to avoid every possible threat in the world (M2-solution): in that case, we would become paranoid and yet probably still fail to imagine something.

To better protect ourselves, one of the strategies has been that of scaffolding our disgust and fear dispositions with smart tools and environmental adjustments that augment or reshape our hardwired machinery. When describing M3, Casati (2017) notices that redesigning the way in which information is presented elicits different cognitive systems, and often provides some cognitive advantage – a strategy that he calls *displacement*. To get what he means intuitively, consider that having introduced the word "displacement" in italics is in itself a very simple example of displacement, so as would have been writing DISPLACEMENT in capital letters: the same informational content is conveyed, but different shapes modulate its salience.

The interesting part of displacement, Casati argues, is that it does not require learning. Similarly, when we build artifacts that scaffold our emotional system, we make use of what I dub *emotional displacement* [ED]. In most cases, ED occurs when:

(1) There is a class of stimuli S, which we rationally appraise as qualifying for a response R, and
(2) the response R is usually regulated by some affect program A, but
(3) A is usually not triggered by S, being only sensitive to the set of stimuli S', and thus
(4) by means of some artifact and/or environmental manipulation, S is translated into S' so that it spurs A into producing response R.

A couple of examples should clarify this.

Smoke alarms—Being trapped in a building on fire is one of the many potential threats of modern age. Obviously, given that buildings are relatively new compared to our evolutionary history, our biological machinery did not manage to catch up with this innovation. We thus had to regulate it by M2. But constantly monitoring every room in a large building is obviously beyond our capacities. Thus, by the end of the Nineteenth century we invented smoke detectors: devices that exploit optic or physical cues to detect smoke, thus inferring the presence of an ongoing fire. Smoke detectors are often coupled with an alarm, which gets activated each time the device reports a fire. Such smoke alarms seem pretty good at their job: a recent report by the U.S. *National Fire Protection Association* claims that "the risk of dying in reported home structure fires is 54% lower in homes with working smoke alarms than in homes with no alarms or none that worked" (Ahrens 2019, p. 1). The trick these machines perform is to translate a stimulus that might easily fail to reach our attention—i.e., the presence of smoke—into something that triggers our fear affect programs—i.e. a loud and annoying noise. A similar principle underlies other security devices like the security systems I mention in the introduction. Now, an interesting feature of smoke alarms is that they sometimes *mis*fire (pun intended). According to the report above, smoke alarm owners claim that roughly three on four alarms were triggered by cooking. In that respect, the way in which smoke detectors are designed mimics our M1 system's "better safe than sorry" philosophy in the trade-off between false negatives and false positives: that is, it prudentially favors the latter over the former.

Odorant for methane—In many countries, cooking requires the use of a flammable gas mix, mostly consisting in methane. While being non-toxic, methane is still a very nasty gas: it is highly flammable, having caused uncountable mining accidents. Moreover, it is asphyxiating. Another reason why methane is so dangerous is that it has no detectable color or smell. For these reasons, you definitely do not want to forget the gas knob open. Wisely enough, in order to reduce home accidents, many legislations issue the obligation to add an odorant to the methane employed for domestic uses: most often, that is tert-butylthiol, a highly odorous substance which might causa *nausea* even in low concentration. By means of this addition, a dangerous stimulus that we would not perceive (methane) is turned into an unpleasant stimulus, which we easily learn to appraise as dangerous.

However, while often EDs are employed to enrich the list of triggers of our affect programs, in some cases, which I dub ED*, it may also be employed to *correct* false positives. ED* cases occur when:

(1) There is a class of stimuli S, which we rationally appraise as *not* qualifying for a response R, and
(2) the response R is usually regulated by some affect program A, but
(3) A *is* usually triggered by S, and thus
(4) by means of some artifact and/or environmental manipulation, S is translated into S' so that it *does not* spur A into producing response R.

An example of this is the following:

Drugs sweeteners—As a child, I was happy whenever I had a cough. The reason is pretty straightforward: the expectorant my parents used to give me in these circumstances was blended with sorbitol, a chemical sweetener, thus resulting into a sweet syrup that I loved. Sweeteners are effective ways to circumvent the typical disgust reaction for the bad tasting chemical substances we sometimes need to employ. Other than in drugs, they are also found in toothpaste: indeed, most toothpastes are mixed with xylitol, conferring them a sweet and fresh taste, without which brushing our teeth would be a much less pleasant operation.[8]

The examples discussed above presume benevolence, i.e., that ED/ED* are employed for a good purpose. Alas, in many cases, ED/ED* can be used with malicious intent, or at least they can produce harmful results even if they were not meant to cause harm in the first place. Such harmful (ab)uses of ED/ED* can be easily spotted in advertisements that elicit our gluttony and vanity to make us buy some products, and even more so in political propaganda that elicits our fear and anger to make us vote for some authoritarian party.

2.4 M4: Offloaded Actions

Neither in M2 nor in M3 our hardwired emotional machinery gets fully sidestepped: rather, it gets—so to speak—'hijacked', that is, its interaction with the environment is mediated by other cognitive systems or environmental scaffolds. By contrast, M4 implies that some external systems replace our affect programs, i.e. that the corresponding tasks are completely offloaded to an external device. This is exactly what the home security system commercial quoted promises to do: to carry the burden of our fear of intruders violating our home.

Or take the case of the smoke alarm discussed above. When a smoke detector triggers an alarm, it does not eliminate the fire threat by itself: what it does is alerting human beings about a dangerous stimulus which might go unnoticed, so that they

[8]While in my example I mostly focused on 'hacking' the input side of affect programs, it is worth mentioning that similar interventions can also affect the output side. The effectiveness of some emotional response might be strengthened or inhibited by means of scaffolds: for instance, if we conceive anger as a means to overcome a rivalry, weapons of all sort can be conceptualized as anger-empowering tools. On the opposite side, handcuffs can be conceived as means to constrain the anger of some potentially dangerous person (or, in other cases, to arouse her/him: the way of emotions are tortuous indeed). See Viola (under review).

can deal with it. But sometimes smoke detectors are coupled with sprinklers, which get automatically activated in cases of fire threats. In this case, the sprinkles take care of the threat by themselves, potentially without any human involvement: we are in M4.

Indeed, technological advances offer several instances of M4 fear-vicarious systems. Smart-home technologies, which are becoming particularly common in the assistance of the elderly and other vulnerable groups of people, include a mix of M3 and M4 systems. For instance, "[s]afety monitors [that] detect environmental hazards such as fires or gas leaks and consequently trigger alerts [and s]ecurity monitors [that] detect and alert to human threats such as intruders" (Demiris and Hensel 2008, p. 33). However, older applications also exist: to remain in the home safety domain, covering electric wires with plastic is a M4 solution that allows us to get rid of the fear of electrocution—i.e., a stimulus that may be evolutionarily difficult to categorize for fear learning. Moving from fear to anger, hence from fleeing from threats to neutralize them, the progressive automation of warfare might represent a shift from M3 to M4.

A more controversial example concerns jealousy. While many researchers overlook the emotion of jealousy, or consider it a mere social construct, Buss (2013) advocated an evolutionary origin and relevance of sexual jealousy, whose adaptive function would be to maximize reproductive success by enacting behaviors aimed at preventing other individuals to have sexual intercourse with one's partner. If that is the case, chastity belts would be a quintessentially M4 sexual-jealousy-vicarious tools—although it is controversial that they were actually used.

3 Implications

Recently, several philosophers have begun to investigate how our emotions extend into the environment (for a review see Krueger and Szanto 2016). For instance, Piredda (2019) talks about 'emotive artifacts' to indicate tools that we invest of some affective value, stressing their relevance in the construction of identity. However, most researchers focused on the feeling aspects of emotions, and thus only took into account artifacts conceived as "tools for feelings" (Slaby 2014, p. 44). As I suggest elsewhere (Viola under review), some attention should also be paid to the emotion-environment interactions with respect to the *function* that emotions play.

The list of systems devised to scaffold and replace emotions (M3 and M4, respectively) is probably very long, and will only get longer as technology advances. The implications brought about by these systems are too far ranging to be discussed exhaustively. For instance, the mechanization of warfare mentioned above raises several cogent moral issues. But less obvious issues might also arise: just like cognitive artifacts working in M4 might endanger the cognitive capacity that they are meant to substitute (Fasoli 2018), artifacts that vicariate our emotions might render us *unprepared* when it comes to actually exercise and manage them. This does not necessarily imply a bad outcome: we might be just fine living in a society where anger gets atrophied. Nonetheless, the least we ought to do is being aware of our reliance on artifacts and tools. While they are devised to overcome the limits of our innate machinery, such devices have their own limitations. To say it jokingly: were crusaders actually employing chastity belts (which seems historically contentious), they might have succeeded in protecting the chastity of their loved ones, but only until the latter did not come across a skilled blacksmith.

A final consideration is that the four modes listed by Casati leave out at least one further important strategy to implement cognitive tasks. Sometimes a task (or an important part thereof) is offloaded not to some cognitive artifact, but to another cognitive agent. Claiming that this M5 strategy is pervasive in our lives would be an understatement: we somehow offload the task of our anger (i.e., to neutralize harms) to police forces and soldiers. We spend public money to pay people who are in charge of predicting and preventing harms that we should seek to avoid, thus doing the job that our fear affect program cannot do all on its own. Some of us spend private money for nutritionist, who hopefully do a better job than our M1 disgust system in advising what to eat. It is worth stressing that this social division of emotional labor need not be accomplished by a human agent: long before home security systems existed, people trained watchdogs to look after themselves and their properties.

Sometimes these socially extended emotion-like processes are coupled with artifacts. For instance, some smoke detectors are not coupled with an alarm or a sprinkler: instead, they automatically call the local firefighters station, so that the agents in charge of this task may take care of the problem.

Delving deeper into the details of the M5 system is clearly a task for another day. In this paper, I content myself with having shed some light on the continuity between the biological evolution of emotions and the technological evolutions of the systems that scaffold or replace their role in survival and reproduction. After having done so, we may be able to better comprehend the nature of some of our interactions with the technological environment we live in, and thus to use and redesign it with due care.

Acknowledgements I am indebted to Colin Klein, Irene Papa, Giulia Piredda, Agostino Pinna Pintor and Sarah Songhorian for their comments on previous versions of this paper. This version also benefitted from Marco Fasoli and Valetina Petrolini's insightful comments. Last but not least, I sincerely thank Tommaso W. Bertolotti for organizing the workshop and putting up this volume.

References

Adolphs R, Anderson DJ (2018) The neuroscience of emotion: a new synthesis. Princeton University Press, Princeton

Ahrens M (2019) Smoke alarms in U.S. home fires. Available at https://www.nfpa.org/-/media/Files/News-and-Research/Fire-statistics-and-reports/Detection-and-signaling/ossmokealarms.pdf

Barrett LF (2017) The theory of constructed emotion: an active inference account of interoception and categorization. Soc Cogn Affect neurosci 12(1):1–23

Buss DM (2013) Sex Jealousy Psychol Topi 22(2):155–182

Casati R (2017) Two, then four modes of functioning of the mind: Towards a unification of "dual" theories of reasoning and theories of cognitive artifacts. In: Zacks J, Taylor H (eds) Representations in mind and world. Routledge, Oxon and New York, pp 7–23
Celeghin A, de Gelder B, Tamietto M (2015) From affective blindsight to emotional consciousness. Conscious Cogn 36:414–425
Clark JA (2010) Relations of homology between higher cognitive emotions and basic emotions. Biol Philos 25(1):75–94
Clark A, Chalmers D (1998) The extended mind. Analysis 58(1):7–19
Damasio AR (1994) Descartes' error: emotion, reason, and the human brain. Putnam Publishing, New York
Demiris G, Hensel BK (2008) Technologies for an aging society: a systematic review of "smart home" applications. Yearb Med Inform 17(01):33–40
Ekman P (1992) An argument for basic emotions. Cogn Emot 6(3–4):169–200
Fasoli M (2018) Substitutive, complementary and constitutive cognitive artifacts: developing an interaction-centered approach. Rev Philos Psychol 9(3):671–687
Faucher L, Tappolet C (2006) Introduction: modularity and the nature of emotions. Can J Philos 36(5):vii–xxxi
Griffiths PE (1997) What emotions really are: the problem of psychological categories. University of Chicago Press, Chicago
Izard CE (2011) Forms and functions of emotions: matters of emotion–cognition interactions. Emot Rev 3(4):371–378
Kahneman D (2011) Thinking, fast and slow. Macmillan
Krueger J, Szanto T (2016) Extended emotions. Philos Compass 11(12):863–878
Laland KN, Odling-Smee J, Feldman MW (2000) Niche construction, biological evolution, and cultural change. Behav Brain Sci 23(1):131–146
LeDoux J (1998) The emotional brain: the mysterious underpinnings of emotional life. Simon and Schuster, New York
LeDoux J (2012) Rethinking the emotional brain. Neuron 73(4):653–676
Lench HC, Bench SW, Darbor KE, Moore M (2015) A functionalist manifesto: Goal-related emotions from an evolutionary perspective. Emot Rev 7(1):90–98
Lindquist KA, Wager TD, Kober H, Bliss-Moreau E, Barrett LF (2012) The brain basis of emotion: a meta-analytic review. Behav Brain sci 35(3):121–143
Norman DA (1991) Cognitive artifacts. Designing Interact Psychol Hum comput Interface 1:17–38
Öhman A, Mineka S (2001) Fears, phobias, and preparedness: toward an evolved module of fear and fear learning. Psychol Rev 108(3):483–522
Olsson A, Ebert JP, Banaji MR, Phelps EA (2005) The role of social groups in the persistence of learned fear. Science 309(5735):785–787
Ortony A, Turner TJ (1990) What's basic about basic emotions? Psychol Rev 97(3):315–331
Panksepp J (1998) Affective neuroscience: the foundations of human and animal emotions. Oxford University Press, New York
Pessoa L, Adolphs R (2010) Emotion processing and the amygdala: from a'low road'to'many roads' of evaluating biological significance. Nat Rev Neurosci 11(11):773
Piredda G (2019) What is an affective artifact? a further development in situated affectivity. Phenomenol Cogn Sci 1–19
Rozin P, Haidt J, McCauley CR (2008) Disgust. In: Lewis M, Haviland-Jones JM, Barrett LF (eds). The handbook of emotions, 3rd edn. Guilford Press, New York and London, (pp 757–776)
Scarantino A (2016) The philosophy of emotions and its impact on affective science. In: Barrett LF, Lewis M, Haviland-Jones JM (eds) Handbook of emotions, 4th edn. Guilford, New York, pp 3–48
Slaby J (2014) Emotions and the extended mind. In: von Scheve C, Salmela M (eds). Collective emotions. Perspectives from psychology, and sociology. Oxford, Oxford University Press, pp 32–46

Stanovich KE, West RF (2000) Individual differences in reasoning: implications for the rationality debate? Behav Brain Sci 23(5):645–665
Sterelny K (2010) Minds: extended or scaffolded? Phenomenol Cogn Sci 9(4):465–481
Susskind JM, Lee DH, Cusi A, Feima R, Grabski W, Anderson AK (2008) Expressing fear enhances sensory acquisition. Nat Neurosci 11(7):843–zzz
Tooby J, Cosmides L (2008) The evolutionary psychology of the emotions and their relationship to internal regulatory variables. In: Lewis M, Haviland-Jones JM, Barrett LF (eds). The handbook of emotions, 3rd edn. Guilford Press, New York and London, (pp 114–137)
Viola M (under review) Three varieties of emotional artifacts
Vytal K, Hamann S (2010) Neuroimaging support for discrete neural correlates of basic emotions: a voxel-based meta-analysis. J Cogn Neurosci 22(12):2864–2885

The manufacturer's authorised representative in the EU is Springer
Nature Customer Service Centre GmbH, Europaplatz 3, 69115 Heidelberg,
Germany. If you have any concerns regarding our products, please
contact ProductSafety@springernature.com

Printed and bound by CPI Group (UK) Ltd, Croydon, CR0 4YY
30/09/2026
02225513-0002